Flow Hive's Book of

Bees and Beekeeping

Cedar and Stuart Anderson

murdoch books

Sydney | London

Contents

Imagine a garden. It could be next to a farmhouse, or in a small backyard, or even on a rooftop in a city. It's spring time – the scent of flowers is on the air, and the birds are singing extra sweetly. Droplets from an early rain shower glisten on bright green leaves. Beneath the ground, earthworms are busy tunnelling through the loamy soil.

In the corner of the garden sits a small wooden house, about the size of a bedside table, with a peaked roof. If you venture closer, you'll hear a gentle humming. Bees are flying around its entrance, their wings glinting in the sunlight as they set off in search of blossoms or return heavy with nectar and brightly coloured pollen bundles. This little home is a Flow Hive. Although only recently invented, it's part of the long and fascinating history of beekeeping.

This is its story – a story that all of us are part of, as beekeeping and humanity are inextricably intertwined.

It's March 2015, autumn in Canberra. I'm bleary-eyed from lack of sleep, nervous and excited. My dad, Stu, and I are standing in my grandfather's garden with a gaggle of reporters and cameras assembled in front of us. After hundreds of hours of work and dozens of failures over nearly as many years, we're about to show the world our invention.

We've been on a quest to create the beekeeper's dream – a way to get honey out of the hive without opening it or disturbing and squashing the bees. Up until now our efforts have run on the smell of an oily rag (quite literally – we fuel our cars with used oil from the local fish and chip shop). We've kept the invention top secret, even from most of our family and friends. Now, that's all about to change.

In the past few days, since we posted a teaser video online, we've been receiving a thousand emails a day. There's a buzz that we can't quite get our heads around. Some people are telling us that we'll save the world, others are curious and seem to be resentful for some reason.

We've never been in front of the press before. One of the cameramen helpfully points out that I've buttoned up my shirt incorrectly. As well as revealing the new invention, we're here to launch a crowdfunding campaign to help manufacture it. The target is US$70,000, which sounds like a lot of money for someone who's never had a new pair of shoes in their life. I've been up all night frantically creating a new crowdfunding page following a last-minute mix-up behind the scenes. My partner, Kylie, heavily pregnant with our first child, is a thousand kilometres away in Byron Bay, making the final edits before we go live. There's a lot riding on this moment. If it doesn't work out, the shed in the woods near the ocean where Kylie and I live will be sold out from under us. We'll have to leave that special place, and the invention my dad and I have worked so hard on won't see the light of day.

Over the past decade, through all the discoveries and dead ends and lightbulb moments and restarts, I've always believed it could be done. But I've often wondered, 'Will anyone even want it?' It's time to find out: here goes …

The cameras roll. I turn the key on the beehive to start the honey flowing. At the same time, Kylie hits 'publish' on the crowdfunding page. Seven minutes later, someone yells, 'You've hit target!'

Forgetting all about the reporters in front of me, I pull out my phone to check the total. Hardly able to believe what I'm seeing, I walk straight out of shot and into Stu's interview to tell him. After eight more minutes, it's reached US$250,000. Two hours later we've broken the record for the fastest crowdfunding campaign to reach US$1 million.

My family in the background is jumping around saying, 'Stop it, stop it, we've got more orders than we can handle, turn it off!', and I'm going, 'Isn't this the kind of thing that dreams are made of?' Looking at the numbers going up and up feels surreal. Most of all, I'm just elated that so many people around the world seem to care about bees and are willing to support the concept of the Flow Hive. Things are in motion now. I don't realise it in the moment, but life has just switched up several gears.

Cedar

Scan here for videos relating to this chapter

Beginnings

How it all started

With the hugely successful launch of the Flow Hive in 2015, suddenly we had more money than we'd ever seen in our lives, and the stress and pressure of fulfilling thousands of orders without any previous knowledge of, or experience in, mass production and manufacturing, all within a nine-month deadline, lay ahead of us. So, how on earth did we end up here?

To an outsider, it may have seemed like an overnight success. But in our own ways, both of us had been shaped by a life of creating and experimenting, making do with what we had, always striving to see how things could work better. This curiosity eventually led us to a discovery that was years in the making.

Our family had always been interested in bees, starting with Stu's dad back in the 1960s. When Stu moved to rural New South Wales and had a young family, keeping bees became part of the bigger picture of a sustainable lifestyle. A sense of wonder and fascination about bees was passed on to the whole family, as we experienced the joy of keeping bees and harvesting honey.

Bees and honey were not our only passions though. Living in a largely self-sufficient community, there was always something to do, to build, to fix, to solve. We loved making things and tinkering, always looking to improve on our ideas. Our home was full of weird and sometimes not so weird inventions, mainly dreamed up by Cedar. And it was one of these ideas that eventually became the basis of what is now the Flow Hive.

STUART ____ ## A barefoot beekeeper

The first time I met a beekeeper, he wasn't wearing any shoes. Not only was he producing his own honey on the outskirts of Melbourne, he was also clearly someone who did things his own way. As a child, this made a huge impression on me, and it must have made an impression on my father, too (the beekeeping, not the shoes), because it wasn't long before we too had our very own beehive. The honey was delicious, and the intrigue, the magic of 'having' a colony of bees, lit me up as well. It helped stoke a passion for the natural world that has been with me all my life.

As kids, we were always encouraged to give things a go, to fix things rather than buy something new, to experiment and try to make things better. I probably inherited the inventing gene from my Scottish grandfather, whose water-powered clothes hoist us kids would play on in the backyard.

When you grow up in an inventive family, there's always an air of questioning. 'How could that be done differently?' 'What do I know and what do I need to find out?' 'What can I use to make this better?'

We moved to Canberra when I was thirteen and it was there that my brothers and I attempted our first bee-related invention: a pedal-powered centrifugal honey extractor using 20 litre oil cans, bicycle parts and bits of wood. We put full frames

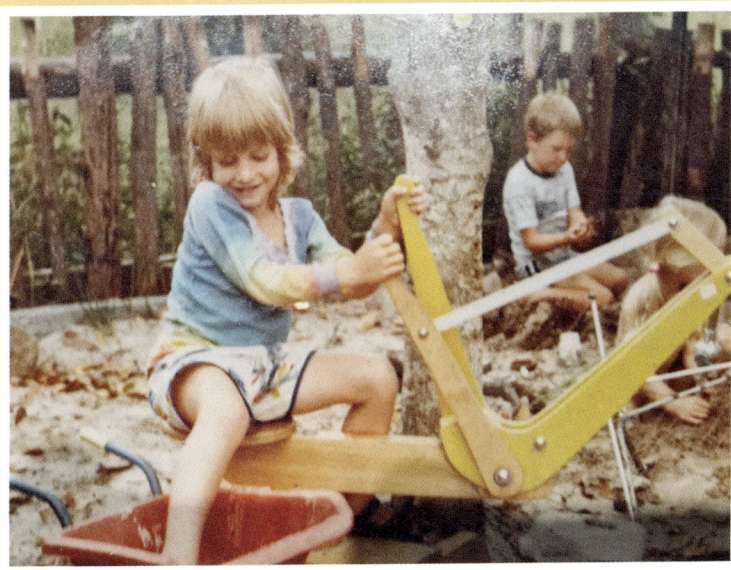

From top Four-year-old Cedar on the tools: a maker in the making; the Anderson brothers, around 1964: Graham, Malcolm, Patrick, Stuart; Cedar on a toy digger, playing with friends.

From top Cedar always loved tinkering and making things, like this motorised go-kart, which made the trip to school quicker; Cedar and Stuart at The Channon, NSW, around 2000.

of honey in the drums and cranked it up. Within a minute we realised that we'd misjudged the forces involved – our contraption tore itself to pieces and left a large amount of honey dripping down the walls of the shed!

When I left home, I travelled overseas, dabbled in university and then, at age 20, headed for a rural life. My partner Radha had a share in an 'intentional community' in the Northern Rivers area of New South Wales: a beautiful region that was starting to attract surfers and hippies looking for a different way of life.

STUART   Life in community

Our new home was 160 hilly acres of beautiful bush full of birds and other wildlife. A waterfall poured out of the rainforest into a creek that fed a crystal-clear swimming hole. People were living in caravans and tents – it was basic, very much a work in progress. I loved everything about it.

I'd always enjoyed doing things with my hands, and being part of a collective project that required practical skills was very rewarding. I was a jack of all trades, helping to establish a water reticulation system and setting up a micro-grid to supply power to the homes in the community. On 'the farm' as we all called it, we helped build each other's houses, contributed knowledge and time to a variety of projects, planted gardens together, and gathered for a big meal every Sunday.

When our local forest looked like it was going to be destroyed by logging we carpooled to the momentous environmental protest at Terania Creek. As we carved out our own way of living and the times they were a-changing, there was a feeling in the air that we might be at the forefront of some sort of new paradigm that would lead the way for humanity. When our non-hierarchical decision-making process led to excruciatingly long meetings every month, we started to wonder how right we'd gotten it. Ah, the idealistic ways of youth!

In March 1978 our twin boys, Chris and Gabe, were born, with Cedar arriving two years later, and my daughter, Mitra, in another four years. I started building the family home out of mud bricks when the kids were small, and this process lasted well into their teens. I was also busy off the farm, going back to university and starting a career in community work. I began working with young offenders, men's groups and environmental awareness programs, and later went on to become the manager of the 'Men & Family Centre'.

The farm was a wonderful environment for our children, and they were always urged to explore and follow their imaginations. When they weren't at school, they roamed from house to house raiding pantries, exploring the rainforest, helping in the vegetable garden and creating their own fun. They built musical instruments, restarted old engines, and even made their own fireworks. We encouraged the kids to learn and not worry too much about making mistakes. Wherever possible, we would allow the consequences of their actions to impact them: if they wanted to stay up past bedtime and then fell asleep on the floor, that's where they'd stay – with a blanket tossed over them.

The kids were all creative, but Cedar often took things to another level. A massive dreamer as a kid, he had a natural bent towards making things and spent his childhood creating, playing and building. We didn't have a TV, and on the weekends, Cedar would usually head to the workshop. When he was around ten, he designed and built his own remote-control aircraft, and it actually flew! We always kept a couple of beehives, mainly as part of a general leaning towards self-sufficiency. Cedar's always really loved honey and he caught the beekeeping bug early; from the age of six he was pottering around the hives with me.

CEDAR _____ # A free-range childhood

I had a completely free-range upbringing and I loved it. Us kids really were 'raised by a village'. I was always making something. My bed had high sides, which was useful as it safely held the mess of my creations – circuit boards, wires and bits and pieces of things I was working on.

When we got a bit over the two-hour walk to school, a few of us kids hit the workshop and built a go-kart using bicycle parts, a motor from an old generator, some gears and the back part of an old mower. The steep rough track down the mountain made a great short cut, though we did end up crashing into the lantana quite spectacularly one day when the brake (a welding wire that pulled a couple of floorboards against the wheels) snapped.

Every now and then, decked out in washing-up gloves and homemade veils that my grandmother had sewn for us, my brothers and I would fill a smoker with gum leaves and set out to raid the beehives. The aim of the game was to get in, get as much honey as we could, and get out with as few stings as possible. I remember once we tried it when the weather was wet, and it turned out that the bees weren't so keen on being disturbed on such a rainy day. We came home with plenty of stings, but it was all worth it for the delicious honey!

My dad's motto was 'get out of the way', meaning let kids be kids, let them learn for themselves and don't try to control them too much. My parents placed a lot of trust in our ability to think for ourselves and didn't criticise us. So, in a way, we grew up fast because we had responsibility for ourselves and others from a young age. Hats off to both my parents for giving my siblings and me that experience. Now that I have children of my own, I know that not criticising is actually a lot harder than you might think.

The evolution of beekeeping

Honey has always been a highly valued food source, and it's widely assumed that humans have been gathering it since long before records began. The earliest known method was to find a wild honey beehive in a tree, pluck up the courage to climb up there and rip it apart, grab as much honeycomb as you can and get out of there. Some Indigenous people still harvest in the wild like this today, and my grandfather used to gather honey in this way too, using a pillowcase to carry the bounty home. Over time, people across the world developed ways to 'keep' honey bees – transporting colonies in logs and eventually building hives from materials like clay, wicker or wood. Hieroglyphs from ancient Egypt feature depictions of cylindrical beehives made of clay and stacked in rows, while the Greeks and Romans made woven hives called 'skeps'. During medieval times in Europe, beehives were often found in monasteries, sometimes built into alcoves in stone walls. Most modern beekeeping is done using hives made from rectangular wooden boxes.

Cedar, seven, and his brother Chris, nine, in homemade protective gear.

A spark of an idea

In his early twenties, Cedar was living with his partner Kylie in a shed on some land by the coast in northern New South Wales. He had a job as a paragliding instructor but it didn't pay much, so to make a bit of extra cash he got 30 hives from a local beekeeper. That was the start of managing a small-scale commercial apiary and selling honey to the local shop.

Extracting honey in the conventional way was always a labour of love – or, some would say, a lot of hot, sweaty, heavy, messy, stingy, hard work. And the bees didn't much like the process either.

First, you have to don your bee suit and gloves. Then fire up a smoker and heft the hives to see which ones are full enough. Chisel off the lids, usually breaking a whole lot of the bees' comb and making a big sticky mess. Pull out the wooden honey frames, trying not to squash too many bees. Brush the bees off the frames, or use a leaf blower. Put those frames into spare boxes and lift them onto the back of the truck, which is not fun when each box can end up weighing something like 30 kilograms (66 lbs). Take them to the processing shed (which in Cedar's case happened to be the shed he lived in). Heat up a knife and start the long process of cutting the wax capping off both sides of every frame, put those frames into a hand-powered centrifuge to spin all the honey out, and then sieve all the dead bees and bits of beeswax out of the honey. After that you have to clean up the big sticky mess and allow the honey to settle to get out all of the wax and bee bits. And as if all of that isn't enough hard work, you've got to go back to the hives, get into your sweaty bee suit, fire up the smoker again, open the hive and put all those frames back where they came from.

From left Cedar and his brother Chris beekeeping; Cedar, 'a born optimist', with his homemade go-kart.

One evening after a long, back-breaking day of harvesting honey, Cedar said to me, 'There has to be a better way. I've had an idea that we could just tap off the honey without opening the hive and disturbing the bees.'

At the time I assumed we were about to enjoy the merry mind-dance we had done so often – one of us makes a proposal, the other comes up with a whacky solution which is topped by a more outrageous idea, then after fifteen minutes or so of satisfying mid-air inventing, the topic changes and that particular conundrum is put aside for the next. This time, however, it was different. He was serious. In fact, he'd already started testing a few ideas.

At first, the idea of removing honey from a hive without disturbing the bees seemed impossible. Imagine a filing cabinet with files hanging in their folders. Now try to figure out how to get a document out without opening the drawer. But we weren't inclined to let a little thing like impossibility get in the way of an exciting idea. Although we didn't know it at the time, this had just sparked a process of inventing that would take a decade to complete – and change our lives.

We weren't inclined to let a little thing like impossibility get in the way ...

CEDAR ____ # Trial and error

I'm a born optimist. *Believing* that you can actually create something is the first step towards making it work. When I seriously started thinking about how to get the honey out of a hive without opening it, one of the first ideas I had was to pull the back off the hexagonal honeycomb cells inside the hive. I thought this might allow the honey to drain down and out, but when I tested it, the honey didn't budge at all. Next, I tried pulling the back and front off the honeycomb. That didn't work either – the surface tension and viscosity of the ripe honey meant it still wouldn't flow.

Then I thought maybe the honey could be sucked out with a tube. I drilled holes in the back of some wax foundation sheets (more on these on page 88) and inserted some tubing to create a vacuum. But that didn't work either. (A few years later, we found a Spanish patent from 1940 for pulling the back off the cells and another patent for sucking honey out with a vacuum pump. Perhaps it's lucky we didn't find those patents early on, because it might have discouraged us from persevering.)

Every time I wanted to try out a new idea, I had to give the bees time to store honey in the prototype (if they even did), so it often took weeks or months to see if a concept could work or not. My dad was very encouraging, as usual, and naturally joined in with the task. Something special happens between us when we work together. I guess our minds are wired in a similar way – we bounce ideas off each other and often only a word, or half a pencil stroke, is enough to make one of us understand what the other is dreaming up. (Our family never lets us play Pictionary together at Christmas because we guess it after drawing just a couple of lines.)

I tried tipping the hive on its side. And that still didn't work. Next, I cast some hexagonal-shaped piston plungers out of epoxy resin to push the honey out of each cell, but – surprise – that also didn't work. After many more variations on these themes, the workshop bench began to overflow with failed experiments.

One morning, when I was half awake, a thought bubbled up. 'Oh, perhaps it can be honeycomb cells when the bees fill it, and then it transforms into something else to drain the honey out.' I started scribbling different shapes, trying to come up with something that would fit together before being pulled apart again. Then, using a craft knife and simple tools, I spent weeks cutting different shapes of comb to construct a hexagonal matrix that could pull apart horizontally. I mounted them using screws and bits of wood and set up an inner tube from an old car tyre as a diaphragm that you could suck on to make the comb come apart. This particular test was the first one that showed promise. I pulled the comb out of the hive and mounted it on the bench, then sucked on the tube and watched the comb come apart. The honey started to drain out slowly. I felt a jolt of excitement. 'Aha, this is getting somewhere.'

CEDAR

Going all in

Up until this point, the whole thing had really been a part-time endeavour. Stu was managing a community organisation focusing on men's health and fathering, while I, as well as teaching paragliding, was doing things like travelling to Sumatra for Greenpeace, spotting illegal burning of forests to help save orangutans from the palm oil industry. The invention was a passion project that I worked on when I had spare time. But now that there was an idea that was starting to look like it could succeed, we both decided to give up virtually all our paid work (and working on other inventions!) so that we could collaborate full-time on this one.

The next breakthrough came one day as we were sitting at the kitchen table in the shed where Kylie and I lived. (It was hard to tell where the workshop ended and the kitchen started – a synergy that suited me, Kylie perhaps not so much, though she was always very supportive and tolerant of my mess!) Stu suddenly put down his coffee, looked up at me and said: 'Hang on a minute, how about instead of pulling the frame apart horizontally, we do it vertically?' As he spoke he gestured in the air, and straight away I understood exactly what he meant. We immediately rushed to the workbench and started building a new prototype.

When we pulled it from the hive a month or two later, we split the cells vertically and watched the honey flow out beautifully into a measuring flask. This design removed the surface tension that had prevented the honey from flowing in earlier efforts, and it was clear there was potential for it to work very well. We designed a full prototype frame and got it 3D printed. This technology was relatively new and it cost us a few thousand dollars to get it done – thankfully Stu's dad, Don, was able to lend us the cash. It was a significant amount of money for two guys without proper jobs or any spare cash, but we knew that it was time to go for it. Once the prototype had been made, we installed it in one of our hives. Then we waited.

Finally, once the bees had filled the comb and capped the cells with wax, it was time to test our invention. We had flown my sister Mirabai (Mira for short) up from Melbourne to film the great occasion. Cameras rolling, we turned the spanner on the outside of the hive which drove a camshaft that operated the parts to split the cells and … snap! The 3D-printed camshaft had broken. Back to the workshop.

We had high hopes that this design would be successful. But the question remained in the back of our minds – 'What if nobody even wants this thing? What if we're just a pair of crackpots tinkering away to no end?'

A week later there we were again, on a sunny afternoon in the paddock with the beehives, this time a metal camshaft in our hands. Feeling hopeful, but cautiously so after having experienced so many failures over so many years, we turned the spanner and watched. I think we were both literally holding our breath. The honey began to flow. Through the observation window we could see it pouring out of the comb and through a tube that ran outside the hive into a jar – the very first jar of honey ever to be extracted without opening the hive and disturbing the bees.

It's hard to describe the feeling of seeing an idea that's been simmering for so long and you've worked so hard on just finally spring to life like that. Our invention worked! The honey was pouring out of the hive while the bees were still standing on the comb surface. (In the Resources section on page 276 you can find a link to a YouTube clip of that first successful test, and hear the excitement in our voices!).

After a few moments of complete euphoria, we sat in the paddock laughing and looking at each other, both thinking the same thing: 'We've done it!'

Now what? Little did we know, in that moment, that the adventure was only beginning, and there were still many challenges ahead. We'd end up having to think outside the box to bring our creation to the world.

The first successful Flow Frame harvest.

Plotting a path to launch

After that first successful test harvest, the elation took a while to subside. Now that we knew the invention could work, we had to get serious about planning a way forward. Dad and I wondered again how people would react to it. If even a small percentage of beekeepers ended up being interested, it could have a lot of potential. We also pondered its impact, because there are plenty of real-world examples of innovations that have unintentionally caused more harm than good. But we were sure that this way of harvesting would be much gentler on the bees than conventional methods, which was surely a good thing. We felt confident we could take our new invention to market ourselves – mainly, as it turns out, because we didn't really understand just how difficult that would be!

The next thing we needed was a patent, so that we'd have the final say on the invention's commercial use and protect our idea from copycats (or so we thought ...). Understanding the patenting process was a steep learning curve and we used up a lot of time finding the right people to help us, but we eventually got there. Our first patent was in 2012, titled 'Improvements to Apiculture'.

Although we knew that the invention worked in our own hives, we needed others to try it out, too, so that we could test assumptions and discover flaws. Stu's father, Don, helped to pay for a plastic injection mould, and when our newly produced frames were ready, we shipped some to beekeepers in the USA and Canada to trial them.

Not all the responses were what we'd expected. An Australian commercial operator watched a demonstration and remained distinctly unimpressed, muttering, 'Oh, yeah, they might like it in Europe.' Pretty devastating to hear. But other responses gave us heart – renowned American beekeeper Michael Bush said, 'This is revolutionary! It's this over-simplistic idea of how a hive can work, yet you've actually accomplished it. That's amazing.'

Choosing to crowdfund

Neither Stu nor I had a dollar to our name at this stage, and we were still unsure about how to launch. We visited other inventors and pored over every case study we could find. People in the business world told us about ways to raise money, but I kept coming back to the idea of crowdfunding. If we allowed people to pre-order the new product online, maybe it could raise enough funds for production and logistics, while also allowing us to connect directly with our customers. One of the most appealing aspects of this strategy was that we'd be able to launch without having to find external investors and we could maintain full autonomy over how we do business.

We went to a crowdfunding workshop hosted by an accomplished inventor to learn more, but when we raised our idea with the expert panel they were discouraging. The first response was, 'Don't do it, it won't work!' Their consensus seemed to be that crowdfunding was for small gadgets, not a complex bulky agriculture product. As we were leaving the workshop, I said to Stu, 'Stuff it, we're going to do it anyway.'

After doing some cost estimations, we set our campaign target at US$70,000: a significant amount of money in our eyes, but we felt like we had a decent chance of getting there. As the path to launch took shape, our excitement grew. We continued to keep the invention a secret from all but a few family members and friends. With this mysterious project taking up more and more of our lives, people who didn't know the details were starting to wonder what the hell we were up to. But we still had to keep it under wraps, because if anything got out our chances of getting a patent could be ruined.

Once we decided to launch via a crowdfunding campaign, we knew that it would be essential to make a great video to show people what we'd created and drum up some enthusiasm. Luckily, my sister Mirabai had studied film-making, so we flew her up whenever we thought we were ready to do another honey extraction test. Often, we'd be all set up to shoot and something would break, or the weather wasn't good, or the bees wouldn't cooperate. It ended up taking more than a year to get all the footage we needed.

CEDAR

Things start heating up

There's nothing like a deadline to get you motivated. Ours was the news that my partner Kylie was pregnant. Ever the optimist, I said, 'Well, we'll just get the crowdfunding thing out of the way before the baby arrives.' The vague deadline we'd set ourselves suddenly narrowed to less than nine months.

We'll just get the crowdfunding thing out of the way before the baby arrives.

Hot on the heels of that wonderful news, a new challenge arose. The owners of the place that Kylie and I called home announced that they were putting the property on the market, so soon we'd have to move out. We loved the location – we'd started our orchards and our apiary was thriving – and this was not a good time to move. I said to Kylie, 'Don't worry, the invention is going to come good,' and asked the property owners if they'd be willing to hold off for just a few more months while I worked on something that would hopefully allow us to make them an offer.

I showed them a video of that first successful test of our invention. They were supportive and encouraging but said, 'We're still going to sell the place.' I guess they found it hard to believe that the couple living so frugally in their ramshackle shed and driving a truck that smelled like chip oil would possibly be able to come up with the money for a piece of land near the coast close to Byron Bay.

Time was running out. We set the date of the launch for February 2015 and started arranging media coverage– now we'd lit a fire under ourselves! With so much to do in the lead-up to the launch, we needed help. Luckily, we had knowledgeable people around us. Saadi Allen, our good friend who happened to be a marketing whiz, helped us to set up a Flow Hive page on Facebook. We also needed a website, so we asked another local friend, Yari McGauley, to build it for us. Before we knew it, we'd grown from a father-son duo into a small team.

It was time to drum up some anticipation! Sitting among piles of prototypes, I pushed enough workshop bits and pieces aside to make room for my old Linux laptop. I logged in to our Facebook page and posted Mirabai's teaser video, which basically said, 'We've invented a way to get the honey out without having to disturb the bees or open the hive, and it's a world first. If you want to find out more, add your email here.' Little did we know a digital avalanche was about to ensue.

Our aim had been to reach 1000 likes and collect 1000 email addresses before the launch. Within 30 hours of the teaser video dropping, we'd had a million views.

I stayed up all night watching a flurry of people sharing their excited opinions – some earnestly telling us that we'd save the world, others saying that we would ruin it. A thousand emails landed in our inbox, and the next day there were another thousand emails. This is when it dawned on us that we weren't actually going to be able to write back to all these people, and we were going to need a proper team. I now realise how lucky we were to have such talented family and friends who piled on board. Everyone was pitching in, even my grandfather. A week later, 70,000 people had signed up to the email list wanting to know more. At that point, although we were stoked to see so much interest, we barely dared to hope that a viral video would translate into a lot of people actually wanting to buy our invention.

CEDAR _____ A sticky situation and last-minute changes

Having assumed it would be difficult to get journalists to come to our place in the forest, we'd decided to hold the public launch in a city instead. Stu's dad, Don, kept bees in the heart of Canberra, so we lined up the ABC (Australia's national broadcaster) to come and film us pressing 'go' on the crowdfunding campaign at his place. Since the teaser video had gone viral, other media started to take notice too.

The only problem was, my grandfather's bees hadn't yet made enough honey for a live harvesting demonstration. So Stu went to one of our hives and pulled off a box packed with Flow Frames full of honey and put it in a suitcase. Getting a box of honey onto a plane wasn't something we had anticipated. At the airport, they said, 'Sorry, your luggage is a kilo too heavy.' Thinking quickly, we ditched the suitcase and checked in just the wooden box of frames. You could see honey dripping through cracks as it moved along the conveyor belt. We crossed our sticky fingers, wondering what the chances were of it turning up at the other end. Thankfully, incredibly, there it was when we landed, a gooey box waiting for us on the luggage carousel. The airport staff didn't say a word.

A few hours later, we put that box of frames onto Don's hive on his balcony in Yarralumla. With the launch scheduled for the next day, a few last-minute decisions still needed to be made. Like, what price is this Flow Hive going to be? Which is actually hard to work out when you haven't made any and, naively, have no idea of the costs of running a company.

In the midst of it all, because of a miscommunication in our little team it was announced to the world that we were switching platforms. So we immediately had to switch from one crowdfunding website to another. I stayed up all night trying to shift everything across, build the page, work out payment systems and so on. The sun came up and I was still there – exhausted, running on adrenaline. The launch was scheduled for 11 am, and I hadn't slept at all. Then the media started to arrive.

Preparing for the launch in 2015. **Clockwise from top left** Stuart and Cedar with a Flow Frame; delivery of the first set of Flow Frames via paraglider; Kylie, Cedar, Jarli, Stu and Michele.

The moment of truth

Back home in the shed where it all began, Kylie, Yari, Saadi and another friend, Terri, bottle of bubbly at the ready, were working frantically on the last bits of the crowdfunding page. Mirabai was uploading the video from Melbourne. The 11 am launch time came and went, but we still didn't have the page ready. Meanwhile, thousands and thousands of people around the world were pressing the refresh button, waiting for us to launch. So many of them were refreshing the page that the Indiegogo crowdfunding platform crashed, and we hadn't even launched yet!

An hour later, we were ready. By now it had been a ten-year-long journey of discoveries and disappointments, fun and effort, driven all the while by a dream of some sort of success. We pretended to press go on the crowdfunding campaign on our phone; meanwhile, back home, Kylie pressed the real start button. What happened next blew our minds.

Seven minutes later we'd blown through our $70,000 target. In seven minutes! We started frantically communicating with the team, putting up more offers on the page because they were running out faster than we could put them up (there's a link to some footage of the launch day in the Resources section on page 276).

The whirlwind begins

The campaign took on a life of its own. Within fifteen minutes, it had attracted US$250,000 in pre-orders from all over the world. Two hours in, we broke the record for the fastest crowdfunding campaign to reach US$1 million.

Even though Stu and my uncles were saying we should stop taking orders, on we went, taking pre-orders just as fast as we could add new batches to the webpage. We were creating new delivery date estimates on the fly according to our best guess as to when we could get manufacturing going and actually deliver to customers in just three months. Seven days later we were declared the biggest crowdfunding campaign ever launched outside the USA. And the orders just kept coming.

The number of enquiries we received over the next six weeks was extreme. Our team quickly grew from just a couple of us to almost 40 people, without a job ad ever being posted. Friends were hiring friends just to help with this mammoth task of responding to what was now thousands of emails every day. The media had gotten hold of the story and it had taken off massively.

There were so many requests for interviews that we simply couldn't respond to, let alone manage, all of them. 'Hello? Oh, no, sorry, I'm too busy. Wait, who? *The New York Times*? Erm, could you possibly come tomorrow morning instead? We've just got to tidy up the shed first. Thanks.' My day would start at 5 am live on a radio station, then often we'd head off to a city for a TV appearance. Meanwhile, I was trying to learn how to make decisions while having to manage people for the first time in my life and trying to work out how to dial up a global supply chain. The team were all pitching in long hours to get everything done.

There were endless comments across social media. People were saying, 'You're going to save the world. You're saving the bees.' Others were saying, 'No, this is terrible, you're going to ruin the world. It's the end of civilisation.' We were going, 'Wow, wow, settle down everyone, it won't save the world. But we can all work together on that. And yes, you still need to look after your bees as you always have done.' There were endless questions like, 'Does it really work?', 'When will my hive ship?', 'Is this a scam?', 'I live on a bus, can I put a hive on the back if I only drive at night?', 'Will you marry me?'.

In the midst of all this chaos, there were late-night conversations on what more could be done to make everything run more smoothly and even what we could do for the planet. Saadi had an idea to raise funds for charity – we raffled a special edition custom-built Flow Hive, and in three days had raised $100,000 for disaster relief in cyclone-hit Vanuatu. A month later we did the same thing for earthquake survivors in Nepal. These were key moments for us because we realised that with a global community we could quickly make a difference.

Although the plan had been to finish the crowdfunding before Kylie gave birth, baby Jarli had other ideas – he decided to come a couple of weeks early. So I said, 'Hang on everyone, wait a sec, I've just got to go and have a baby.' Back in the shed, Kylie's waters had broken and she was in labour. We called our midwife friend who came and let us know it wasn't looking good for a home birth. We jumped in the back of her LandCruiser so she could take us to hospital an hour away.

Kylie said, 'I have to keep my bum in the air to stop the baby coming, but I need to work on dilation, so you'll have to kiss me.' (Apparently, kissing releases oxytocin in the body, which aids in dilation.) So there we were in the back of the troop carrier, our heads pressed to the floor and bums in the air, kissing as we drove down the winding roads. The thought came to me that my life had now been literally turned upside down.

When I got back on deck, we'd become the biggest crowdfund in Indiegogo's history at that time. By the time the eight-week campaign closed, we'd raised US$12.2 million in total and received almost 25,000 pre-orders from customers in more than 130 countries. The level of support and encouragement we'd received was astonishing. So many people were feeling inspired by our invention and the story behind it.

We were amazed, exhausted and happy. It had taken ten years of work to become an 'overnight success'. Our new adventure was just beginning.

Building the Flow Hive

CEDAR

How are we going to make all these hives?

Looking back, the crowdfund feels like a honeymoon period, because once it was over we found ourselves with a three-month deadline to produce and deliver almost 25,000 Flow Hives. More orders were coming in every day, and we had no clear path to making any of it happen. I'd always been a very relaxed person, but for the first time ever I could understand what it was like to feel anxious. In the past we'd worked on one problem after another, but now we had to sort out manufacturing, supplier agreements, quality control, accounting, logistics, global export arrangements and the necessary legal intricacies – all at the same time! Before the launch, my dad and I had talked about what a relief it would be to not have to keep our project a secret anymore, but rather than solace we were mostly feeling stress.

That's when the hard work really started.

We found a manufacturer in Brisbane that was willing to work with us to make the Flow Frames as sustainably as possible, and after dealing with a couple of technical issues they were able to hit full steam with a 24/7 manufacturing and assembly line – something pretty rare in Australia. At the same time, we contracted a company in Oregon to make the wooden components.

Pressure mounted as the delivery deadline for the first pre-orders came closer and closer. We really wanted to be the first major crowdfunding campaign to deliver on time. So, when we saw that we were falling behind, everybody worked flat out to make it happen. Stu reckons he pretty much left his body for that whole period. When the first Flow Frames finally came off the production line, just for fun I delivered them to the winner of our Vanuatu fundraiser raffle – in person, by paraglider, dressed in a bee costume. I'm pretty sure it was the world's first drone bee package delivery!

It felt really great to get hives to just about everybody who had ordered during the campaign, in time for Christmas. And interest stayed high after the launch, so much so that we couldn't make hives quickly enough to meet demand. It was a great problem to have, even if we didn't have much time to stop and appreciate it.

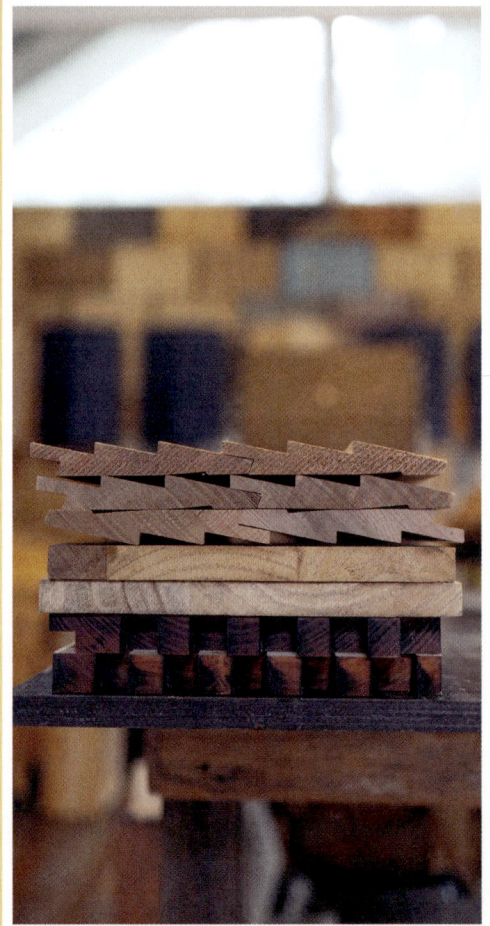

Clockwise from top left Rear window cover latches; Cedar in the factory during the Flow Hive 2 launch; roof and brood box components; CAD drawings on the workbench.

A crash course in entrepreneurship

Long before the crowdfund, in what now seemed like another life, experts had advised us to come up with a business plan. More than once we started tentatively writing things up on a whiteboard, then one of us would have an idea about the invention and we'd be gone. The business plan never got finished. We've come to learn that trying to reinvent the world at every step can leave you going nowhere fast. (In our heads we've already come up with solutions to half the world's problems, though our infinitely patient partners would probably prefer we gave a little more time to things like fixing the toilet that still doesn't flush properly.) Thankfully, in the early days of the company we brought in my sister-in-law, Summer Cook, and she soon became our general manager. She's our queen bee, an indispensable asset in so many ways.

Everyone was working very long hours during the early days, and inevitably there were moments when someone got upset. When something goes wrong and we lose a bit of money, that doesn't bother us much, it's just part of running a business. But if you've got an upset friend, that will keep you up at night. Whenever a meeting started to get too tense, I'd reach for a box of wigs and costumes that we kept in the cupboard, and once everybody was sitting around the table looking a bit silly the mood would usually lighten.

It seemed like the pace would never let up. The motivation for inventing the thing in the first place was so that I'd have to work less, not more. (Some say that necessity is the mother of invention, but others say it's laziness.) It was ironic that to make something easier, you'd end up spending a decade of your life figuring it out and then working absolutely flat out to make everything come to fruition. At the same time as all this was going on, I had a new baby at home. I've honestly never been more overwhelmed in my life. And there were these business people getting in my ear saying, 'You have to sell it. There's no way you'll be able to make this all work. Sell it now while you can.'

It sounded like reasonable advice. But what would someone else do with our invention? How could we be sure that a different owner wouldn't do something that was bad for the bees or the planet? We felt like we couldn't possibly sell out, that we owed it to these people all over the world – who had believed in us and given us their hard-earned cash – to do right by them by delivering the best hive we could. And, as hectic as the journey was, our problem-solving brains loved the challenge. Over time, more and more elements fell into place and the manic vibe transformed into something more stable. Today, Flow's HQ sits on an escarpment overlooking the coastal heath and out to the ocean, where you can see humpback whales breaching in the distance as they pass by each year. When the wind blows just right I can take off with my paraglider from out the front to fly across the valley alongside the wedge-tailed eagles (a great way to see what's in flower). Much of the food in our daily shared lunch comes from a garden bed that we eat sitting next to. There are beehives everywhere. We've built a natural swimming pool, there's mead bubbling in the basement, and of course the shed is full of tools and experiments.

We have a wonderful team in place doing everything needed to keep the hive going, and every once in a while we'll take the team on a surprise adventure like hot air ballooning, a snorkelling trip on a remote coral reef, or skydiving. We want to keep the feeling of being on that inspired edge, and it's a way of saying thanks to them too.

Towards the end of 2017, we established our own laser-cutting factory to manufacture timber components. We've installed a huge solar array on the roof to turn sunbeams to laser beams. By the time the third anniversary of our original launch rolled around, we finally felt like we were getting a handle on what we were doing – so we figured it must be time for a sequel. In 2018, we dropped a new and improved version of the Flow Hive, again through Indiegogo. Echoing the original launch, the campaign coincided with another even more exciting new arrival: Jarli's little sister, Mella!

Another invention that had been many years in the making followed in 2025 – the Super Lifter. It's a device that makes it much easier to open a beehive. We'll talk about it in more detail on page 133.

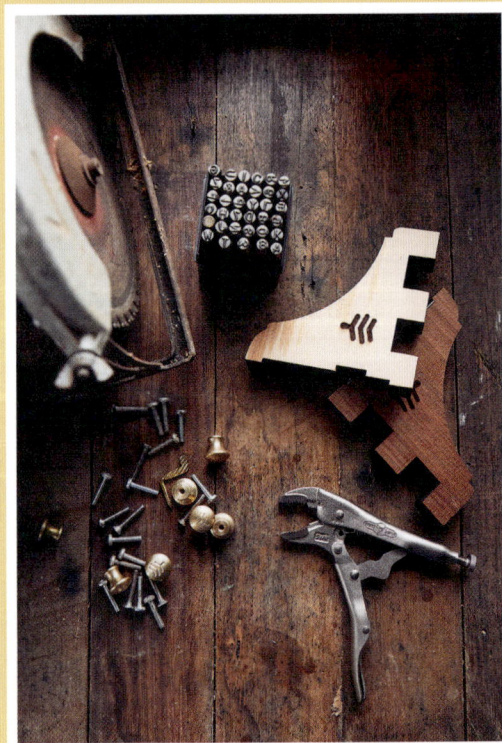

Clockwise from top left Cedar in the workshop; Flow Hive components; the original assembly line. **Opposite page** Stuart and Cedar at work.

Our community and the Flow-on effect

Throughout this whole story, the thing that's blown Stu and me away the most is how many people have supported us and wanted to be a part of what Flow is doing. And not only that, but also how much they fall in love with their bees and start seeing the world a little differently as a result. Looking after bees connects people to nature in a very direct way, and it often inspires them to get curious about the other little beings that are vital to life on our planet.

Beekeeping has always been surrounded by some mystique, and up until recently, it's been mostly associated with older guys. The Flow Hive has made beekeeping a lot simpler to take up as a hobby, and we've tried to demystify it a bit by providing tons of educational resources and an online course (TheBeekeeper. org) so that people can learn how to take good care of their bees. It's been fantastic to see a wider range of people becoming beekeepers.

We hear a lot of touching stories from customers about how beekeeping has changed their lives. It's no wonder that five-year-olds and ninety-year-olds and celebrities have fallen in love with the art. You end up with more honey than you can eat, it's a great pastime to share with family and friends, your garden gets a pollination boost, and bees are properly fascinating!

Above The Flow team in the Byron Bay office. **Opposite** Cedar, Mella and Jarli harvesting honey while Kylie watches.

What's it all for?

Stu and I are still a bit gobsmacked whenever we take a moment to reflect on how things have changed since those early days in the shed. Our lifestyles are still pretty simple. Although the business requires a lot of time and effort, we're still making things in the shed and coming up with ideas – though of course now we have an amazing team to help us.

And, yes, I was able to get that block of land before the owners sold it. Phew! Now Kylie, Jarli, Mella and I, plus three goats, four echidnas, two koalas, 22 wallabies and 2.4 million bees can all live there happily ever after.

What really keeps my dad and me going now is being able to make a difference in the world. More on this in the Impact chapter (page 259).

Scan here for videos relating to this chapter

1

Bees,
the universe
and everything

> 'The earth laughs in flowers.'
>
> *Hamatreya*, RALPH WALDO EMERSON

Imagine, for a moment, our world without flowering plants.

There would be no apples, no chocolate, no coffee. Forget about the flavours of strawberries, peaches and cherries. The scent of jasmine would never tickle a single nostril, and nobody would experience that feeling you get at the sight of a meadow in bloom.

It's a scary thought to a modern human, right? Especially Stu, who loves his coffee!

Well, a flowerless planet was once the case – roughly 130 million years ago. At that time the Earth's landmass was covered in green vegetation, traversed by huge dinosaurs and our own rodent-like ancestors. Flowers, however, were absent. Because plants can't move around, to reproduce they had to spread their pollen on the wind and hope for the best.

Pollination and co-evolution

Pollen grains are very tiny kaleidoscopic-looking structures that contain genetic information. For two plants to combine their genes, those little granules must be transferred from the male part of one individual plant onto the female part of another. The grains are wrapped in a protective layer of protein, which makes pollen quite nutritious, and this led to certain ancient species of wasps developing a taste for it.

As these wasps and other insects inadvertently carried a few pollen grains from one plant to the next, they caused the reproductive rates of those botanical species to outstrip others. Over a long stretch of time one thing led to another which led to another, and eventually the intricate mechanism of mutation and co-dependence that we call evolution resulted in the emergence of flowers and their vital counterparts – bees.

Bees can be distinguished from wasps primarily by their furry bodies and the fact that they collect nectar. Today there are roughly 20,000 species of bee on Earth and they exist in myriad shapes and colours. There's the tiny dark orange *Perdita minima* that lives beneath the ground in the south-western deserts of the United States and weighs only a third of a milligram. The multicoloured jewel-like orchid bee (from the tribe Euglossini) flits through South American jungles. Australia is home to much-loved species such as the cuddly-looking teddy bear bee (*Amegilla bombiformis*), the neon cuckoo bee (*Thyreus nitidulus*), which has a bright turquoise body and slick black wings, and the blue-banded *Amegilla cingulata*. In Indonesia there's the black 'king of bees', raja ofu (*Megachile pluto*), which can grow to almost the size of a human thumb. And let's not forget our good friend, the world-famous European honey bee (*Apis mellifera*).

A plethora of pollinators

One honey bee can pollinate up to 5000 flowers in a single day. It's estimated that roughly a third of our food depends on pollination by bees. And this crucial function is carried out not only by honey bees or even just by the 20,000 or so other species of bees; there are butterflies and moths and birds and bats and possums and lizards and other creatures that pollinate plants too!

In northern New South Wales where we live, there's a stunning array of blossoming flora. You get explosions of purple when the jacaranda trees bloom in October and grevilleas flourishing in a variety of hues in April, which is our autumn. These colours send a message seeking assistance from other more mobile living things in the dating game. Flowers are where plants keep their reproductive organs, and – as is not uncommon in matters of romance – appeals are made to sensuality. The plant makes itself attractive not to others of its own species, but to the pollinators that will help it mate. Their beauty is in the eye, and the nose, of the beholder.

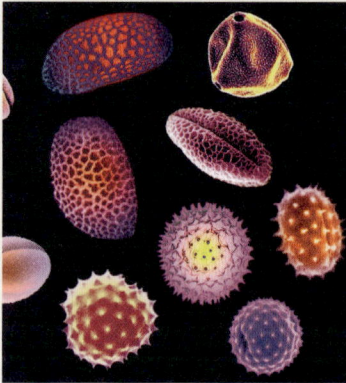

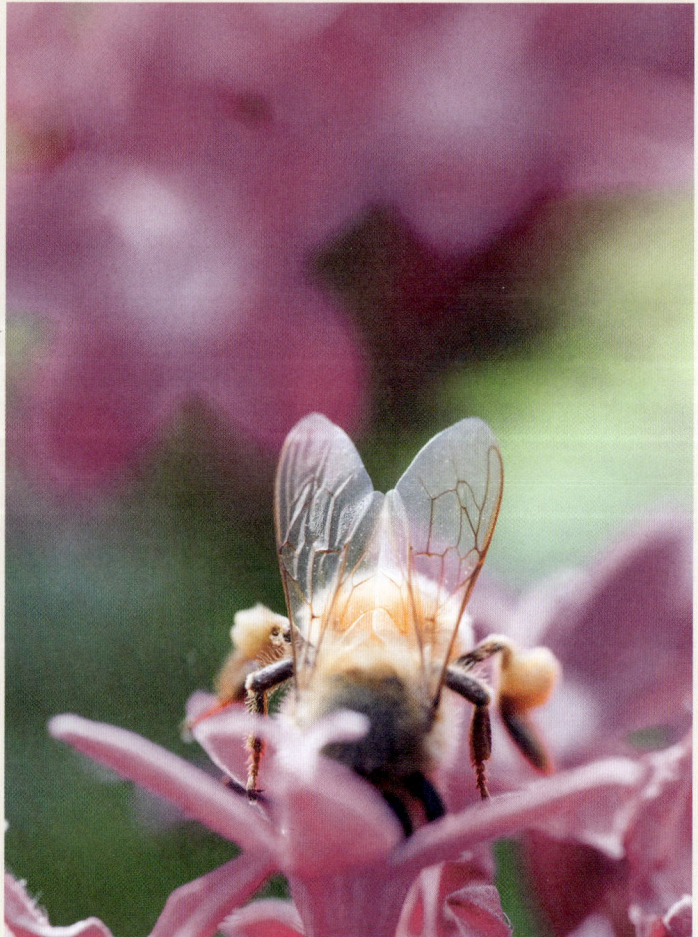

Clockwise from top left Pollen granules under a microscope; a honey bee foraging; arriving back at the hive with pollen baskets loaded.

Bees have five eyes. They can see ultraviolet light and process images five times faster than we do. They also smell phenomenally well, using receptors on their antennae. Flowers have developed their fantastic colours and scents as a way of advertising the location of their nectar stores to passing bees. Some even have arrow-shaped UV guides on their petals, which are invisible to us but to a bee look like vivid bullseyes.

Invisible forces

Each flower produces a unique negatively charged electrical field that is influenced by the flower's shape and height above the ground. When a bee comes close, those electrical fields cause tiny hairs on her body to bend, so she can actually sense flowers that are a few body lengths away from her. The charge can cause pollen to jump from the flower onto the bee before she even lands on it. When she does land, she will leave electrostatic footprints on the flower's petals. The next bee who arrives at that flower will be able to sense this and use this information to help her decide whether to spend time checking it for nectar.

Picture a bee on a foraging trip. She's flown far from her home and she knows that flowers are what she's after. Wooed by the promise of a sugary reward, she lands on one. Her extraordinary sense of touch helps her to recognise the microscopically fine texture of the petals. She goes straight to the flower's nectary, tasting it through receptors on her feet. Using her long feathery tongue, she begins to fill her crop or 'honey stomach'. As she drinks, the branched hairs on her body rub against the flower's stamens (its male part) and she is dusted with pollen. She'll brush this into 'baskets' on the knees of her back legs, allowing her to collect the equivalent of up to a third of her own weight. When she next visits a flower of the same species, some of those pollen grains will rub off from her body onto its stigma (female part). Fertilisation can now occur – the plants have accomplished their aim with the bee as their intermediary, while she's achieved her aim of finding food.

Flight and navigation

Let's leave the flowers for now and stay with our hungry honey bee (to help us talk about how amazing she is, we'll pretend she has a name – Stu reckons we should call her Bea, short for Beatrice). Once Bea has collected enough nectar for one trip, it's time to fly back home. With such small wings compared to her body size, she has to sweep her wings backwards and forwards rapidly in a twisting motion to create spiralling vortices of air that will give her the amount of lift she needs to take off. She'll beat her wings 230 times per second to stay airborne, and to keep her energy up she can tap a little nectar from her honey stomach as she flies. Enzymes in her stomach have already started to break the nectar down into simpler molecules, beginning the process of turning it into honey. If it's been a bountiful foraging expedition, which could mean visiting up to 100 flowers, she could end up carrying almost her own weight in nectar and pollen back to the colony.

To find her way home, she'll use her extraordinarily powerful sight. Each of the distinctive compound eyes on each side of her head has around 6900 facets (lenses), all pointing in different directions, with tiny hairs where each facet meets another. The hairs are thought to be used to gauge the direction of the wind, helping her stay on course on a blustery day, and also possibly to collect pollen. As well as being able to see UV, she can perceive polarised light. This lets her orient to the position of the sun, even on a cloudy day. Bees are also great at distinguishing between dark and light, so can easily recognise edges and identify landmarks.

Besides her vision, Bea has a map-like spatial memory that she hones by undertaking several training flights close to home to memorise important landmarks before she ventures far away. She can also sense the earth's magnetic field. It's theorised that this is accomplished using tiny crystals of a mineral called magnetite in her abdomen which act as magnetoreceptors, a bit like having your own internal compass. Altogether, these sensory abilities allow for tremendous navigational capabilities. Some species of bee can find their way back home after flying as far as 14 kilometres (8.7 miles) away from their nest, which is the equivalent of millions of times their body length.

Home is where the hive (or burrow) is

Bea arrives at her home. What it looks like varies considerably depending on what type of bee she is. Most species are solitary, often living in tunnels they've excavated in mud or wood. Sometimes they line their burrows with leaves, resins or – in the case of certain species – their own self-made natural bioplastic. Mason bees, carpenter bees, orchid bees and cuckoo bees are all solitary.

A small number of species aren't solitary but rather eusocial (from the Greek *eu-* meaning 'good'); they live in sophisticated communal structures and work together to benefit the collective rather than the individual. Stingless eusocial bees such as *Melipona* and *Trigona* often use a mix of plant resins and other materials to build their nests. Bumblebees are also eusocial; they live in burrows and use wax secreted from their bodies to create small pots for storing nectar and pollen.

Bees make bread

Adult forager honey bees don't eat pollen. They stop producing the enzymes necessary to digest it when they graduate to flying out of the hive. After they've brought some back and deposited it in the cells near the developing brood, other workers add a little nectar and pack the pollen in tightly using their heads. Once a cell has been filled, it's sealed with a layer of honey and left to ferment. Microbes cause chemical reactions that make the nutrients in the pollen more readily available when this nutritious 'bee bread' is fed to the developing brood.

The most well-known bees of all, honey bees, are also eusocial and also make wax, which they use in great quantities to sculpt their homes. Let's say Bea is a honey bee. As she lands at the hive entrance, you'll often be able to see colourful lumps of pollen on her hind legs. She passes the nectar she's collected to another worker bee who takes it away to be stored, and then she finds a suitable cell in the brood chamber to deposit her pollen haul into. If she's been scouting and found a particularly good source of nectar while she was out, she'll want to tell her sisters where it is.

Communication

We've both always been fascinated by bees and bee communication. Before we invented the Flow Frames, we used to extract a lot of honey in the conventional way in the shed Cedar lived in. There would always be some interactions with curious bees while we were there. If the door to the shed was open, inevitably at some point a scout bee would come past, detect the honey, and go back to the hive to do a little dance to tell the other bees all about it. It wouldn't take long before a whole lot of foraging bees would show up at the door. Naturally, that would make us shut the door, but because it was so hot, we'd need to open a window. Although the window was only a metre away from the door, the bees would all still be waiting at the door to be allowed in. They had followed their instructions and were sticking to them. Eventually, another scout would come by the window, note the honey and disappear again to the hive to spread the word about the 'new' location. Soon, the forager bees would be at the window. Then we would shut the window and open the door and so on and so on until the honey processing was finished. We found this amazing. How, in a dark hive packed with 50,000 bees, are they able to communicate such accurate information about the window and the door?
We have so much to learn about these furry little insects.

Dance language

The mystery of how honey bees tell each other the location of a good food source puzzled humans for a long time. Aristotle himself wondered about it over 2300 years ago, but it wasn't until 1973 that German-Austrian ethologist Karl von Frisch won a Nobel Prize for decoding the 'waggle dance'. A newly returning bee moves in a repeating figure-of-eight pattern, waggling her body vigorously when she crosses the centre of the pattern. The angle of this part of the dance conveys the direction of the food in relation to where the sun is, and the length of that part of her dance represents the distance from the hive. And the better the nectar source, the more enthusiastically she waggles.

While the dance is being performed (in the dark, remember), other bees stay close, sometimes repeating her movements. They then head out and fly directly to the foraging spot. Their skill is so sophisticated that if a bee needs to repeat the directions later that day to a new group of bees, she will adjust the dance to reflect the new position of the sun. If there's a lot of nectar coming in, foragers will also perform a 'tremble dance' to recruit more receiver bees to collect the bounty as it's brought into the hive.

With their phenomenal ability to communicate through 'dance', honey bees are able to maximise their chances of finding enough food to survive, and can make decisions collectively when finding a new home. (We've tried communicating through dancing in our office, but unfortunately it appears that email has won out in the end.)

They also possess one of the most complex pheromonal communication systems in all of nature. Our forager bee will have encountered a cloud of chemical signals upon her return to the hive. These chemicals cause changes in the bees' physiology and behaviour, coordinating their activities and the social organisation of the colony. For example, pheromones can be used to raise the alarm if something is threatening, and the ratio of nurse bees to forager bees is kept at an ideal balance by a pheromone that the older foragers release which slows the maturation of nurses. You'll sometimes see worker bees standing near the hive entrance with their abdomens raised, their wings fanning vigorously to spread the Nasonov pheromone stored in their bellies, which helps their returning sisters to find their way. It also tells the foragers to come back and stay if the hive needs defending.

Although bees can't hear sounds travelling in the air very well, vibration is also used as a means of communication in their world. In her antennae and her knees, Bea has organs much like ears that can sense tremors travelling through the comb she's standing on. Workers can buzz to express alarm to each other, and queens create vibrations to coordinate the workers' behaviour.

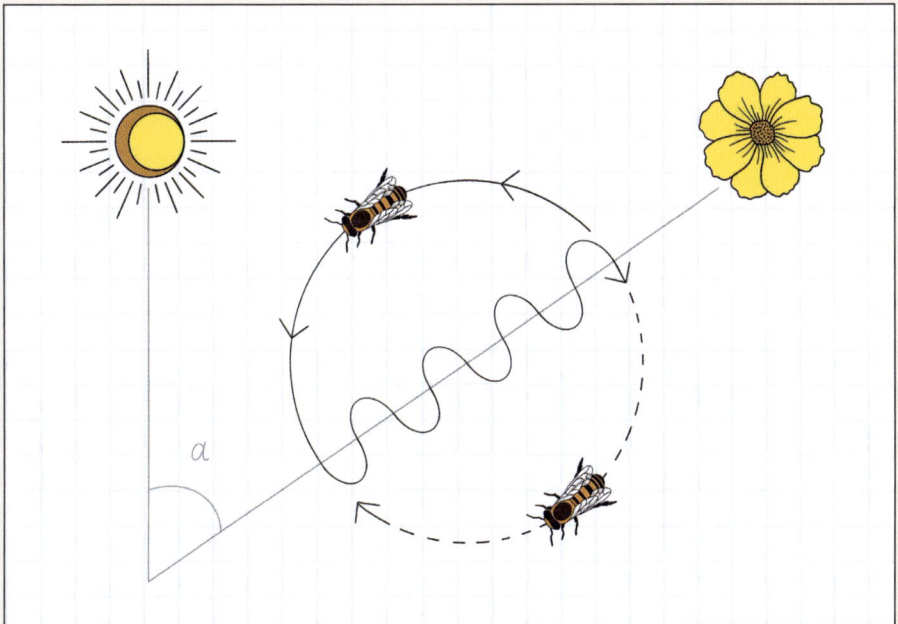

When performing the waggle dance, a bee will waggle her body as she moves in a direction that indicates the angular direction of the food source relative to the sun.

Bee intelligence

It requires a fair bit of brain power for our friend Bea to undertake all these elaborate endeavours. Although her brain contains only a million neurons (we humans have more than 86 billion), these are tightly packed into an organ the size of a sesame seed and form one of the most intricate structures that occurs in the known universe. Scientists who study bee cognition are often astonished by how clever these little creatures turn out to be. As well as their ability to navigate and convey information to each other, bees can measure distance, count, understand the concept of zero and even do basic maths. They're able to recognise images of human faces. They've been trained to play 'golf' and to distinguish between the differing styles of Impressionist painters. They can learn simple tool use, understand abstract concepts, solve problems, pay attention selectively, remember things and teach each other. There's even evidence of bees passing behavioural traits to each other in a cultural fashion, in the form of 'traditions' specific to groups. Brain size isn't everything after all.

Bee neuroscience

Scientists have been carrying out ingenious experiments in their exploration of the bee brain. Discoveries about how it functions could pave the way for important insights into the consciousness of humans and other animals. In Australia, Professor Andrew Barron has created a computational model of a bee's neuroanatomy and says that the bee brain could be 'a model for studying diverse intelligences'. Research in this area also has many potential practical applications in fields like AI and robotics; for example, in the use of robot 'swarms' to assist emergency services in post-disaster situations.

Because the purpose of bees' lives is subsumed into the collective superorganism of the colony, people could be forgiven for thinking of bees as mindless automata with no individuality. However, studies indicate that there's much more going on than just a drive to work for the survival of the collective – bees also have their own preferences and even feelings. One recent experiment showed that worker bees have varying desires to perform specific jobs within the hive – much like our team! Another found that an individual bee can be more or less pessimistic than its sisters.

Professor Lars Chittka, one of the world's foremost bee scientists, believes that bees have 'emotion-like states'. Like us, bees have fluctuating levels of serotonin and dopamine in their brains, and opioid-producing pleasure centres. They learn not only from each other, but also from their own negative or positive experiences, and will show signs of 'trauma' after encountering a predator. They might even dream – some research indicates that 'bees might have dream-like experiences analogous to non-REM dreams in humans'. And this is wild – when placed in an enclosure with a set of colourful wooden balls, without receiving any material reward for doing so, bumblebees have been observed energetically rolling around on them. This is the first known example of insects engaging in play!

We can't know what Bea's inner experience is like or to what degree she might be sentient. It certainly does seem that something very subjective is going on behind those marvellous eyes of hers. And it's exciting to wonder what else will be discovered about her mind in the coming years. We couldn't agree more with Lars Chittka when he says of bees: 'These unique minds, regardless of how much they may differ from our own, have as much justification to exist as we do. It [shows] a wholly new aspect of how weird and wonderful the world is around us.'

The experiment by Professor Lars Chittka and his colleagues showed that bumblebees can create mental imagery, a 'building block of consciousness'.

Everything's connected, man!

Flowers and bees represent a perfect integration of pollination, co-evolution and innate intelligence – it makes your mind boggle at how incredible nature is. The symbiotic relationship between plants and pollinators combines nutrition and reproduction – two of the most fundamental biological drives – to lead to a glorious profusion of living things all over the planet, including ourselves. It's a magnificent illustration of life's basis in mutual interdependence.

To get a deeper sense of this interdependence, it is possible to 'walk in the shoes' of insects or other animals. We are all connected to the whole of life on Earth; it's just that we can be oblivious to this fact. Stu used to co-facilitate a three-day workshop in which the aim was to use a variety of activities to give the participants (including himself) a closer sense of connection to other life forms. One key exercise was 'The Council': the workshop participants would allow different animals, plants or aspects of nature to 'choose' them and then would represent and speak for these beings in a 'council', discussing the situation facing species and ecosystems on Earth. It was a fascinating experience. On the one hand, it seemed pretty silly. Yet at the same time it was profound. Stu definitely had a shift in his understanding of himself. While it felt awkward, there was a genuine experience of being that particular animal, plant or ecosystem. Stu says his sense of 'himself' expanded.

One of the most rewarding things about becoming a beekeeper is that you get to steward a small but very important part of the relationship between plants and pollinators. We can thank Bea and her kin for the fact that we get to taste apples and smell jasmine and we can talk about all the ways we can help them out. As John Muir, one of America's most influential naturalists and conservationists, says, 'When we try to pick out anything by itself, we find it hitched to everything else in the universe.'

Clockwise from top Two globe mallow bees (*Diadasia diminuta*) asleep in a globe mallow flower in the USA; Australian native blue-banded bee, of the genus *Amegilla*; a bee of the genus *Bombus*, commonly known as a bumblebee.

Rooftop beekeeping

Charles, New York, USA

Charles is a mathematics professor whose rooftop apiary is an oasis of calm in bustling Brooklyn. The abundance of different flowers in the city results in consistent forage for the bees and a diverse mix of honey flavours. To see more of Charles's story, check out the Resources section on page 276.

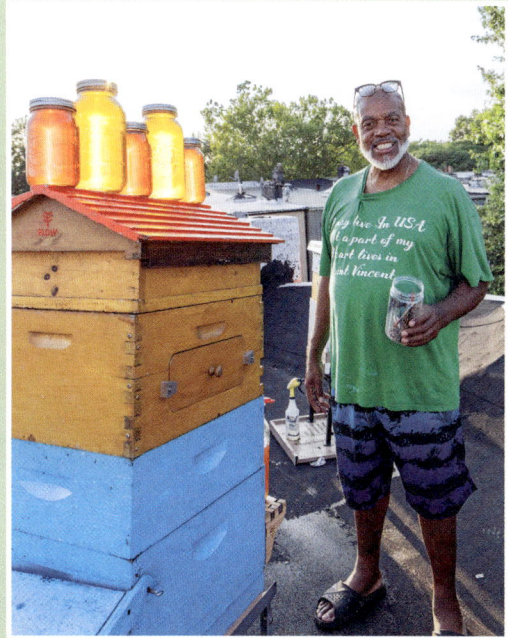

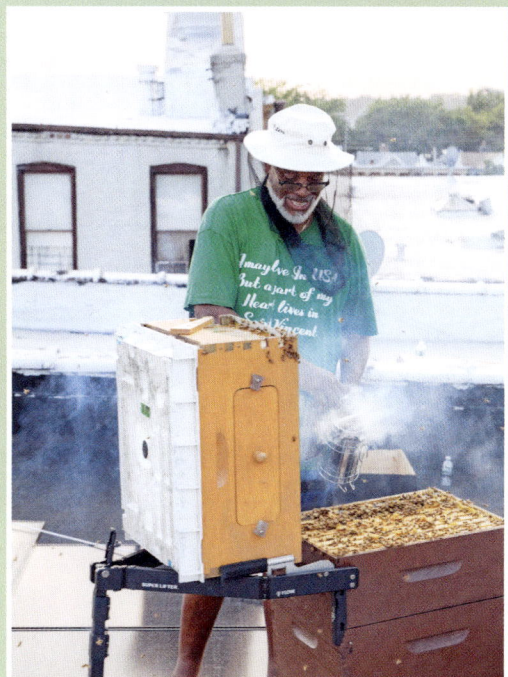

'Treat the bees well and you will be rewarded. When I retire and go back to the Caribbean, Flow Hive is coming with me.'

Scan here for videos relating to this chapter

2

Beekeeping basics

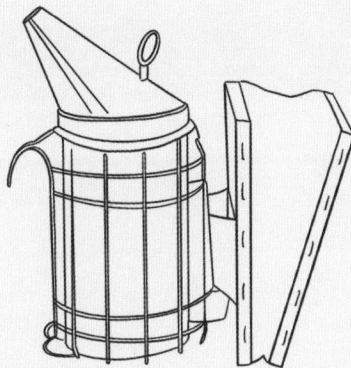

'One can no more approach people without love
than one can approach bees without care.
Such is the quality of bees ...'

LEO TOLSTOY

We originally started keeping bees solely because of the honey. Stu's grandfather had goats and chooks (chickens), and the dinner table often featured homegrown veggies. The goats produced milk too, although young Stu wasn't very keen on how it tasted! Living as we did on the farm in northern New South Wales, a lot of our food came from the garden. Given Stu's prior experience keeping bees, having our own honey was a natural progression.

The longer we kept hives, the more interested we became in the bees themselves – their intelligence, their peculiar behaviours, the way thousands of them could operate as a single organism. And we were also aware that by keeping bees we were taking part in a very ancient practice.

In China, mi-feng (honey bees) appear in art dating back 3000 years. However, the Cuevas de la Araña, a UNESCO World Heritage site near Valencia, Spain, offer a glimpse into an even earlier history of honey gathering. The cave's rock art details the essential rituals of a hunter-gatherer society, including a now famous painting of honey gathering using a rope and basket. This rare insight into early beekeeping is considered somewhere between 8000 and 9000 years old.

An ancient practice

As honey was a high-calorie food and one of the only sources of sweetness in existence, early humans were willing to go to great lengths to gather it, braving stings and dangerous heights. The practice of collecting wild honey using crude climbing equipment and stone hatchets continues within some Indigenous communities around the world to this day.

Collecting honey from wild hives is believed to have been widely practised everywhere honey-producing bees were common: by !Kung in Africa, Australian First Nations people, ancient Mayans in Central America and early civilisations in China, Egypt and Greece. Basic harvesting from wild hives evolved into a hybrid form of beekeeping – bee colonies were moved into logs or other vessels to make honey collection easier. From there, the development of manufactured hives and a more organised system of beekeeping eventually emerged.

In an intriguing example of interspecies cooperation, tribespeople in East Africa have an age-old tradition of communicating with birds who show them where to find wild beehives. These birds love to eat the bees' wax combs, but they can't get to them without help as the hives are difficult to crack open.

The communication goes both ways: when an adult greater honeyguide bird (*Indicator indicator*) finds a hive, it will often use a particular chirp to seek the assistance of nearby humans, and when people are out searching for wild honey they will use a traditional call to summon the bird. Honey hunters learn this call from their fathers and pass it on to their sons. The Yao people in Mozambique use a call that sounds like 'brr-hm' to recruit honeyguides, whereas Hadza honey-hunters in Tanzania make a melodious whistling sound for the same purpose.

Using sounds and gestures, the bird leads the hunter to the location of a hive, where he will smoke out the bees and crack open the nest, taking home the honey and leaving the wax combs and larvae for the birds to feast on. In the folklore of some East African cultures, hunters are encouraged to thank the honeyguide with a gift of some honey, otherwise it may punish its follower in future by leading him to a lion, bull elephant or snake.

In the Middle Ages, Europeans constructed hives from wicker, reed or stone and sometimes built them into alcoves in stone walls. Art depicting beekeepers at work, complete with protective masks and tunics, first appeared in the 16th century. By the 17th and 18th centuries, as a result of trade and colonisation, European-style beekeeping and *Apis mellifera*, the European honey bee, had arrived in the Americas, Australia and many other parts of the world.

Although we know little about early beekeeping practices, it's clear that bees were venerated culturally. There are countless examples of them being mentioned in myths, fairytales, folklore, legends, poetry and music, where they were commonly used to represent the virtues of diligence, cooperation and order.

Clockwise from below Bee hieroglyph from ancient Egypt, where bees were believed to have been created from the tears of the sun god Ra; this ancient rock art from Cuevas de la Araña in Spain is about 8000 years old and depicts honey being gathered from a wild hive. It's the oldest known image of bees and the earliest evidence of honey consumption by humans; *The Beekeepers and the Birdnester*, a drawing by Pieter Bruegel the Elder (c. 1568), shows apiarists with wicker masks and woven 'skep' style hives.

Clockwise from top left Bees on comb;
jars of golden honey after collection; honey
harvest from a Flow Frame; a beekeeper
at work.

As honey has been so significant to humans as a food and medicine throughout the ages, it's not surprising that traditions demonstrating respect for the importance of bees arose and became commonplace. In some parts of Europe and later the United States, a household's bees had to be informed whenever certain critical life events occurred. Specific traditions varied: in some regions, bees would receive a slice of wedding cake when a marriage took place; in others, the newlyweds were obliged to 'introduce' themselves to the hive.

The most important rituals relate to death and persist among some beekeepers to this day. If a beekeeper or their close family member dies, the bees must be informed, lest they abscond or themselves weaken and die. The practice of 'telling the bees' involves knocking gently once on each hive and verbally passing on news of the death, sometimes in rhyme. Hives may also be draped in black fabric for a period of mourning.

Customs like these illustrate the bond that beekeepers feel with bees beyond seeing them solely as livestock.

So if you're curious to participate in this long, rich heritage of beekeeping, we'd be the first to say don't hesitate, give it a go! People who begin keeping bees quite often tell us that they wish they'd started years ago. It's such an extraordinary thing to be involved with as you learn so much about the workings of nature and gain an opportunity to connect with the natural world in a new way. And then there's the honey (and beekeeping stories) you'll be able to share with your family and friends.

We didn't know that much about bees when we started out – useful information wasn't always easy to come by. However, these days it's much easier to get access to knowledge than it used to be. We've got tons of detailed resources online (see page 276), and you can look for a mentor at a local beekeeping club.

The art of keeping bees lies in understanding them: how they behave, their biology, the way that their environment and its cycle of change affects them. It really is an endless journey of learning and fascination. Let's run through some of the main things you'll need to know to get started.

Meet the bees

When people talk about keeping bees, they're usually referring to a single species in the Apidae family: the 'honey bee' or more specifically the European honey bee (*Apis mellifera*). It's important to recognise that this is just one of almost 20,000 species of bees in the world. Species other than honey bees are typically called 'native bees' (even when they've been introduced). Both honey bees and native bees are absolutely crucial to the sustainability of plant life on our planet.

Native bees

As we mentioned in Chapter 1, native bees come in many colours, shapes and sizes. They're essential for pollinating native plants and vary in their hive-making, honey-making and pollinating habits. Most are solitary; some live in colonies and produce small amounts of honey. North America has around 4000 native bee species and Australia has over 3000.

Honey bees

Thanks to their outstanding pollinating skills and the extraordinary amount of honey they make, people have taken honey bees with them just about everywhere they have settled worldwide. They are now the most common type of bee in the world, found on every continent except Antarctica. Often referred to as 'super pollinators', they pollinate more than 70 per cent of the world's main crops. No wonder they're considered essential to human life!

A honey bee colony comprises three types (or 'castes') of bees: the queen, female worker bees and male drones. A single colony can include anywhere from 20,000 to 80,000 bees, with 90 per cent of the hive being workers.

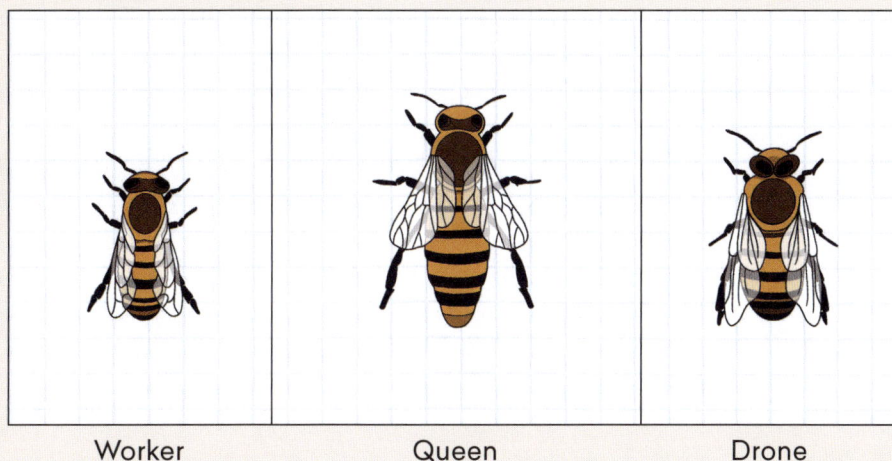

Worker   Queen   Drone

The queen

The queen is the mother of all the bees in the colony. Her genetics (and those of the drones she's mated with) will determine the colony's traits, such as how productive, docile and tolerant to illnesses they are. The queen is the largest bee in the hive, and she's the only one capable of laying fertilised eggs.

Outside of rare temporary circumstances, there is only one queen in a hive at a time. Her primary role is to mate and then lay egg after egg to sustain the population of the colony. Mating happens during her 'nuptial flight' in the first two weeks of her adult life. After that, she spends the rest of her time in the hive laying eggs. Fertilised eggs become the next generation of female workers and unfertilised eggs become male drones. A healthy queen can lay up to 2000 eggs daily, more than her own body weight! Her life usually spans a couple of years, but they have been known to live for five years or sometimes more.

People often think that the queen is the 'ruler' of the hive and the workers are totally subservient to her, but that's not the case at all. If for some reason she weakens and cannot fulfil her egg-laying duty to a satisfactory level, the worker bees will stage a revolution by dethroning her and installing a new monarch. Sometimes they'll even commit regicide by surrounding her in a tight cluster and vibrating their bodies to raise the temperature until she dies. This is called 'balling the queen'.

Queen bee surrounded by her retinue of attendants.

The health and wellbeing of the queen are of the utmost importance to the colony, but if her output lessens because of age or illness, the worker bees will start raising another queen to take her place. Survival of the colony is paramount, and if a queen dies suddenly, another is made as soon as possible. When the hive is preparing to replicate itself by swarming, the bees will make a number of spare queens. The first one to emerge will kill all her rivals.

A queen is created when a single larva in a specially shaped cell is fed exclusively on royal jelly, a highly nutritious substance produced by young worker bees. Larvae destined to become workers or drones receive royal jelly for just three days before switching to 'bee bread' (fermented pollen), while a queen receives a diet of pure royal jelly for her entire life – only the best for her royal highness! Although all honey bee larvae start out genetically the same, a queen's development is 'reprogrammed' by the amount of royal jelly she receives. This food activates genes associated with ovary development, growth, longevity and behavioural changes, while suppressing others that would give her the smaller, sterile body of a worker.

The queen bee also emits pheromones (often through her feet) that play crucial roles in the coordination of the colony, helping to maintain cohesion within the hive by regulating worker behaviours like foraging and nursing. One of these pheromones is produced in her mandibles (jaws) and signals her presence to the worker bees while inhibiting the development of their ovaries and suppressing their instinct to rear new queens. So as well as laying all the eggs, she plays a vital role in the overall organisation of the hive's activities. If only our leaders were this focused on promoting cooperation and harmony among their subjects ...

For info on how to spot the queen when you're beekeeping, see page 142.

Workers

As their name implies, worker bees are busy creatures. Throughout their lives they'll take on multiple different roles. A furry new worker's first task is to clean the cell she just emerged from, and she'll spend the first day or two of her life doing this. Cells are inspected by the queen. If she's not pleased with the state of a cell she won't lay in it and the worker bee must repeat the cleaning process.

The worker moves on from cleaning duties to nursing the new larvae, and after that she'll take on other tasks depending on the colony's needs. Her next endeavour could be anything from tending to the queen to making wax for the comb, receiving and packing food from returning foragers, fanning the hive to regulate its temperature, vibrating to create warmth when it's cold or foraging for nectar and pollen. Foraging is the final role for all workers. After three weeks or so of collecting nectar and pollen, their wings will be torn and their bodies worn out.

However, not all worker bees become foragers; a few of them will spend time as guards, water gatherers or undertakers. Although bees certainly live up to their reputation for being busy, they also need to rest – you will often see them sleeping in a cell or on the comb surface not doing much,

Drones

A drone has a larger body than a worker, no stinger and just one main purpose in life: to meet a queen and mate with her. His large eyes are positioned towards the top of his head so that he can see the queen from underneath against the sky. He'll usually be found at a drone congregation area. Here drones from multiple colonies fly around in a large group, waiting for queens to arrive. When it's time for a virgin queen to make her nuptial flight, she heads to the local drone congregation area.

Drones vastly outnumber queens, so mating competition is fierce and most drones never manage it. The act happens mid-flight, and although this makes for an impressive airborne feat, it's a short-lived one lasting less than five seconds. Having successfully mated, the drone's endophallus (his equivalent of a penis) breaks off with an audible 'pop', and he falls to the ground and dies. The process repeats, and the queen may mate with 30 or so drones. The drones that make it back home are often driven from the hive in winter as the hive conserves its resources, because the drones don't contribute to core colony functions such as gathering food or nursing the brood.

Female worker bee drinking nectar through her proboscis.

Male drone bee: note his larger eyes and abdomen.

The lifecycle of a honey bee

Honey bee eggs hatch into larvae three days after being laid. Nurse bees feed each larva hundreds of times per day over six days. Then the larva's cell is capped over with porous wax, and the larva spins a silk cocoon around itself so its adult body parts can begin to develop. This is known as the pupal stage. It usually takes 21 days for a fertilised egg to develop into a worker bee, while it takes a drone a bit longer – around 24 days. Queens typically spend around eight days in the larval stage and then a further eight days in the pupal stage.

A worker bee will generally work hard for about six weeks over spring or summer, after which time her wings will give out and her life comes to an end. However, workers born just before winter, when the hive's workload significantly reduces, may live up to six months. In contrast, a drone's lifespan is usually only about 55 days.

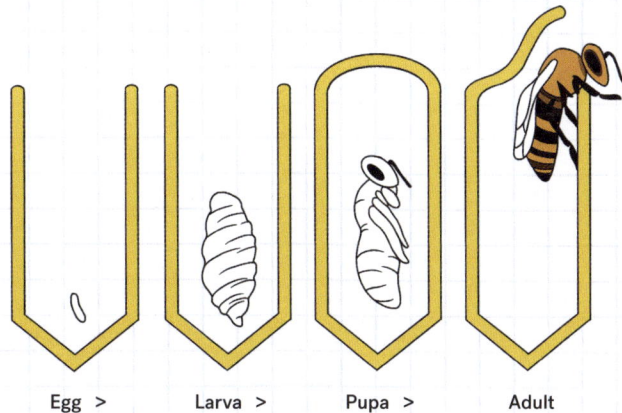

Honey bees develop through four distinct life cycle stages

Egg >   Larva >   Pupa >   Adult

The time to develop from egg to adult differs
depending on the type of bee

	Egg hatches	Larva capped	Adult emerges	
Queen	Day 3	Day 8	Day 16	Begins laying 2–3 weeks after emerging
Worker	Day 3	Day 9	Day 21	
Drone	Day 3	Day 10	Day 24	Flies to drone congregation area around 2 weeks after emerging

Honeybee subspecies

There are roughly 10 different *Apis* species, which are often informally categorised into three groups: macro bees, micro bees, and cavity-nesting bees. All of those bees originate from different parts of Asia, except for *A. mellifera*, which is believed to have originated from Africa or the Middle East and later migrated into Europe. *Apis mellifera* spread across the globe, largely aided by humans, which means they now inhabit a wide range of landscapes and climatic zones.

Different environmental factors – such as temperature, landform, pests, diseases and floral sources – result in bees developing particular traits and behaviours to adapt to their environment. There are now approximately 33 distinct *A. mellifera* subspecies worldwide.

For example, *A. mellifera carnica*, also known as Carniolan bees, originate from the Balkan Peninsula, where temperatures are generally colder. They are often dark in colour, are able to overwinter well, are gentle but can defend themselves vigorously. The Italian honey bee, *A. mellifera ligustica*, on the other hand, originates from Italy and is usually lighter in colour. This subspecies is admired for its calmer temperament and high honey yield.

Depending on the specific traits that a beekeeper requires, different genetics will be more or less suitable. Queen breeders commonly breed for temperament, brood production, pest and disease tolerance and honey production. As long as the quality is good, it's usually better to buy queens local to your environment.

The superorganism

A 'superorganism' is composed of many individuals that act as a single, unified system. In a honey bee colony, each worker has her own role to play but acts for the benefit of the collective, even if it means paying the ultimate price by sacrificing herself to protect the hive! Every bee works towards the colony's goals, which are survival and reproduction (much the same as most individual living creatures).

Communication through pheromones enables the bees to act as a cohesive entity. These chemicals trigger different behaviours in response to changing conditions. And the notion of a 'hive mind' might even have some basis in reality – a 2018 study by the University of Sheffield found that a bee colony, when considered as a whole, behaves similarly to the way a human brain does when required to choose between several options.

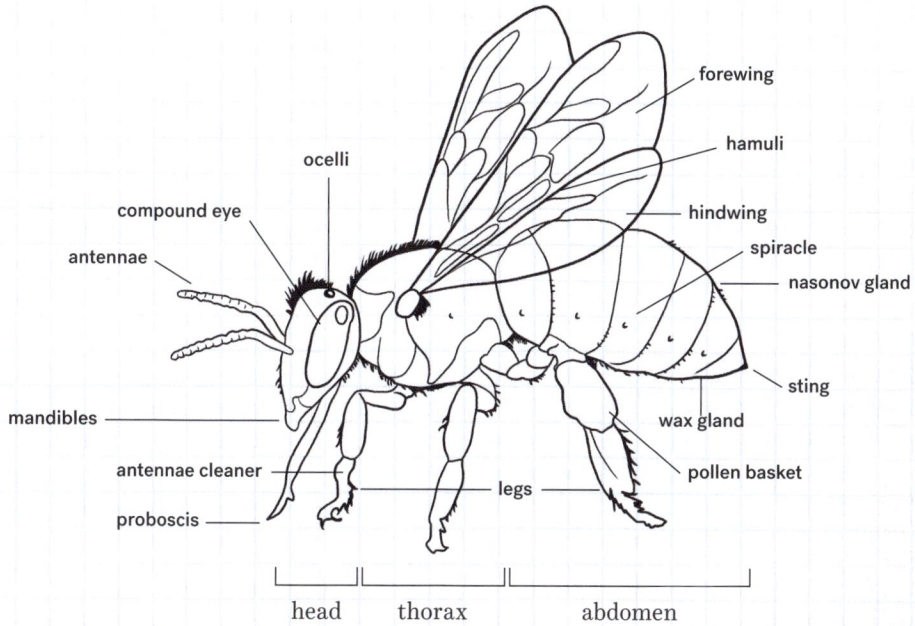

Honey bee
Apis Mellifera

- ocelli
- compound eye
- antennae
- forewing
- hamuli
- hindwing
- spiracle
- nasonov gland
- mandibles
- antennae cleaner
- proboscis
- legs
- sting
- wax gland
- pollen basket

head · thorax · abdomen

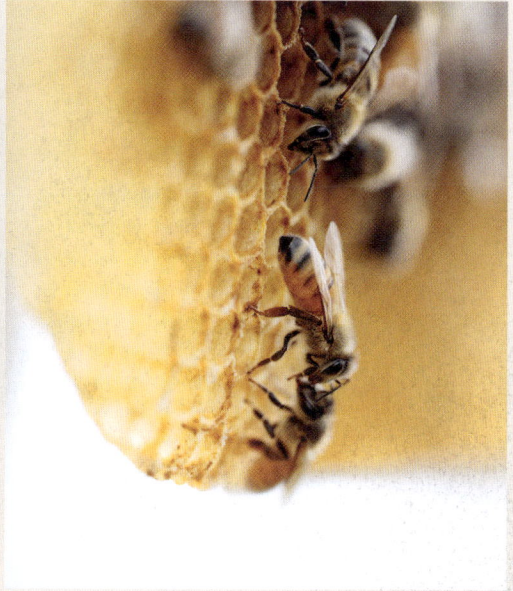

Left A drone emerging from his cell. **Above** These bees could be communicating or passing nectar. **Following pages** A fluffy newly hatched bee (on the right) communicating with one of her sisters.

Bee nutrition

Until they pupate, worker and drone brood are fed royal jelly. After this they are fed a fermented combination of nectar, pollen and honey, called bee bread, which we discuss on page 45. Bee bread comes in various colours and its composition depends on what's in flower locally and its stage of fermentation.

After the pupa has turned into a young worker bee and emerged from its cell, it continues eating bee bread until it is about five days old and is ready to start leaving the hive to forage. It then switches to a diet of nectar and honey, as it can no longer produce the enzymes needed to digest the bee bread.

If nectar supplies are low, bees use other sweet foods, such as ripe fruits or honeydew, as an energy source. Giving your bees sugar water during a low-nectar supply can sustain a hive in lean periods. During the testing process for the Flow Hive frames, for example, we often fed bees sugar water to speed up the bees' honey-making process so we could see if our prototypes were working. The honey produced this way wasn't worth keeping, but it helped us get a faster result.

Go to page 207 for a deeper dive into bee nutrition.

Before you start

There are a few more things it's good to be aware of when you're getting ready to embark on the adventure of beekeeping, and lots that you can do to get ready for the arrival of your new winged friends.

Where can bees be kept?

People keep honey bees in all sorts of environments and climates, from urban rooftops and balconies to rural landscapes, suburban backyards and dry, tropical or cold regions. As long as there are flowering plants around for some of the year, your bees should be able to thrive. There's more info on page 93 about choosing where to put your hive.

When to begin?

Spring is usually the best time of year to get your first hive going. As the Earth wakes up after winter and flowers start to bloom, nectar and pollen are to be found in abundance. Bees use this supply of energy and nutrients to build up their numbers by creating comb and raising new generations of workers and drones. So it's the easiest time to get bees and usually provides the best conditions for them in their new home. (For more on how to source bees for your new hive, see page 100.)

How much space is needed?

A lot of people are surprised by how small a space you can keep bees in, especially with a Flow Hive, as it eliminates the need for harvesting equipment. It's even possible to have bees on a balcony in an apartment building, as long as they have a clear enough flight path and you've got room to conduct inspections. It's important to check local regulations and also talk to the people who live close by.

Availability of forage

Bees can live just about anywhere that flowering plants are present, but nutrition is very important for their health and they really do best with a diet of nectar and pollen from varied sources. If you live in a region where there are few or no flowers for long periods every year, you may need to help your bees at times by giving them additional food. Honey bees are capable of flying over 9.5 kilometres (6 miles) in search of food. They typically prefer to forage within a much shorter radius when possible, as shorter flights conserve energy and time.

Being a good neighbour

Consider the proximity to your neighbours when looking for the best place to put your hive. If they're very close and you think they might have some concerns, you could consider talking to them first. The offer of an occasional jar of honey could be helpful! Talking to people in your community about bees can even end up being beneficial to pollinator health in your area. Many people are keen to know how they can support healthy bee habitats. They're interested in the reasons that pollination is so crucial to our environment, the importance of their gardens and ways to reduce the use of highly hazardous pesticides.

Make sure your bees are of gentle stock if you're keeping them in a location close to other people. Relocate or requeen if the temperament of your hive is showing aggression.

When choosing a location for your beehive (more on this on page 93), be sure to position it away from walkways and don't point the entrance towards anything that might get damaged by bee poo such as cars and washing lines, as the tiny yellow dots are very sticky and can be tricky to remove.

As well as being considerate of your neighbours, ensure you're aware of any local regulations regarding beekeeping. Although rules on keeping bees vary across countries, states and jurisdictions, hive registration is a very common requirement. Beehive registration assists with maintaining biosecurity, so always check with your local authority to determine what's required in your area. (See the Resources section on page 276 for handy links.)

Beekeeping safety

A lot of people's first experience with a bee is being stung by one as a child. Bee stings aren't usually dangerous unless you have an allergy, in which case you need to be really careful around bees, and needless to say, probably shouldn't take up beekeeping.

Although bee stings aren't particularly dangerous for most people, they can hurt and cause swelling and itchiness. Most people prefer not to get stung very often, so please keep yourself safe by using the proper equipment, especially when you're starting out. Protective gear is very effective and ideally you'll be dealing with nice placid bees most of the time.

How and why do bees sting?

A worker bee will sting as a means of defence if she perceives a threat. Although she can sting other insects with thin exoskeletons repeatedly, her stinger is barbed and will lodge in the skin of mammals. If the bee stings a human, her abdomen is often torn from her body as she tries to withdraw the stinger, causing a slow death.

Venom is injected when the stinger is activated. Most people experience a bee sting as a sharp pain that subsides fairly quickly, with swelling and redness that fades after a few hours or sometimes lasts for days. Many people think that swelling around a bee sting indicates that they're allergic to bees, but local swelling is a natural reaction to the bee's venom. It might be painful, but it doesn't indicate an allergy. However, some people can have severe allergic reactions to stings (anaphylaxis). This requires urgent treatment as it can be life-threatening, and it's important to know how to recognise and respond to the symptoms.

Understanding anaphylaxis

Anaphylaxis is an acute, multi-system allergic reaction that can lead to respiratory collapse if not treated urgently. It is a *life-threatening medical emergency*. Signs of anaphylaxis reaction to a bee sting include the following: abdominal pain or vomiting; difficulty breathing or noisy breathing; swollen tongue; swelling or tightness in the throat; a wheeze or persistent cough; difficulty talking or swallowing and/or a hoarse voice; persistent dizziness or collapse; and, in the case of young children, becoming alarmingly pale and floppy. Signs of anaphylaxis can arise up to two hours after the sting.

It's important to be up to date with first aid practices, and be sure to read the info on how to deal with an anaphylaxis reaction at the link in the Resources section on page 276.

Clockwise from below The Flow team in beekeeping suits, worn to prevent stings; safely inspecting a frame; using a smoker to calm the bees. **Following pages** A just-hatched worker bee.

What to do if you're stung

If you're a beekeeper, it's likely that you'll get stung once in a while. If it happens while you're working your bees, simply stay relaxed, remove the stinger, smoke the sting site and the bees and, if you're not too uncomfortable, resume your beekeeping.

To reduce a sting's impact, flick or scrape the stinger out as soon as possible using the edge of your fingernail, your hive tool or a credit card. Try to avoid squeezing the venom sac attached to the stinger as you don't want more venom going into the skin. Once you've removed the stinger, wash the area with water and soap and apply a cold compress. If needed, take some mild pain relief and apply calamine or other lotion to reduce itching. Don't scratch the sting; it can worsen the wound and increase the risk of infection.

When a bee attacks, she releases an alarm pheromone that smells a little like banana to alert her sisters. This pheromone lingers, so if you've been stung while beekeeping, it can help to smoke the area to mask the smell.

Asian giant hornets (*Vespa mandarinia*) are a fearsome predator of honey bees. They are much bigger than bees and their thick exoskeleton shields them from stings. This allows them to enter hives and wreak havoc, decapitating bees with their sharp mandibles. Japanese honey bees (*Apis cerana japonica*) have evolved a defence against this threat which cleverly utilises the fact that their bodies can tolerate temperatures 2 degrees higher than the hornets are capable of withstanding. The bees swarm the attackers and use their wing muscles to generate heat, literally cooking the hornets alive!

Reducing the chance of stings

Although bee stings can happen anywhere bees are active, it's when you're working at or are very close to a hive that the likelihood is greatest. We recommend that you always wear appropriate protective clothing including a bee suit, veil and gloves (see page 97 for details) while working with bees, especially when starting out.

Avoid wearing dark coloured clothing if you're getting close to a hive (it's believed that the bees might perceive you as a bear or other threat). Brightly coloured clothing that could echo the hues of flowers and hence attract bees is also best avoided, as is anything woollen as they could get their legs stuck in it. A protective veil can be worn over your head and face (make sure the veil is not directly touching your face so that bees can't make contact with it, and that any long hair is tucked underneath). Some people also prefer to remove or cover watches, rings and dangling accessories that bees could get caught in.

Bees are extremely sensitive to smell, so if you're getting close to a hive it's best not to wear perfume, strong deodorants, aftershave or hairspray. And if you've been in contact with strong animal or fuel odours it might be preferable to wash and change your clothes before approaching your bees.

Once when Stu was spending some time at an apiary of 40 hives, he noticed that the bees were upset at him, not at the others in the group he was with. He realised that the shampoo he'd used two days before had left a lingering scent that the bees clearly didn't like. He changed products and was no longer singled out for attack.

Another good habit is to not leave open sugary drinks near hives! If you have an open drink, a can of soft drink for example, always check the contents before drinking in case a curious bee has mistaken it for nectar.

Warm, still, sunny days are the ideal conditions for beekeeping. Definitely don't work your bees on a very rainy day, as they don't like this sort of weather and it can make them grumpy and more aggressive. Always move gently when opening a hive to help keep them calm. If you're a beginner, bring your mobile phone with you in case you need to dial an expert, and it's preferable to have someone experienced give you a hand the first couple of times you get into the hive. They can advise you on how to handle frames and give you tips as you go.

The next steps

Now that you've gotten to know the fundamentals of a honey bee colony, let's take a tour of a beehive. We'll show you how it's structured, where to situate it, how to suit up properly and, most exciting of all, how to get and install your bees!

Mirabai with her hive on a Berlin balcony.

MEET THE BEEKEEPER

Beekeeping brings peace

Jonny Richter, New South Wales, Australia

With the help of his daughter Storm, Jonny Richter runs five Flow Hives and sells the honey on a roadside stall and to local cafés. But even when harvesting 40–60 kilograms (90–130 lbs) per hive each year, it's hard to keep up with demand! To see more of Johnny's story, check out the Resources section on page 276.

'When I come home from beekeeping, I'm always calm and mellow. The bees give so much to me and my family, so I've got to treat them well and they treat me well in return.'

Scan here for videos relating to this chapter

3

Beehive basics

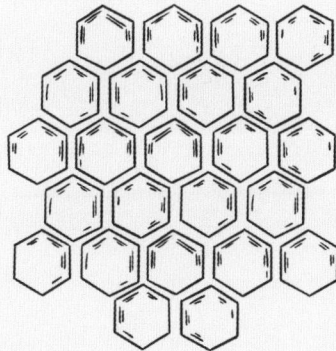

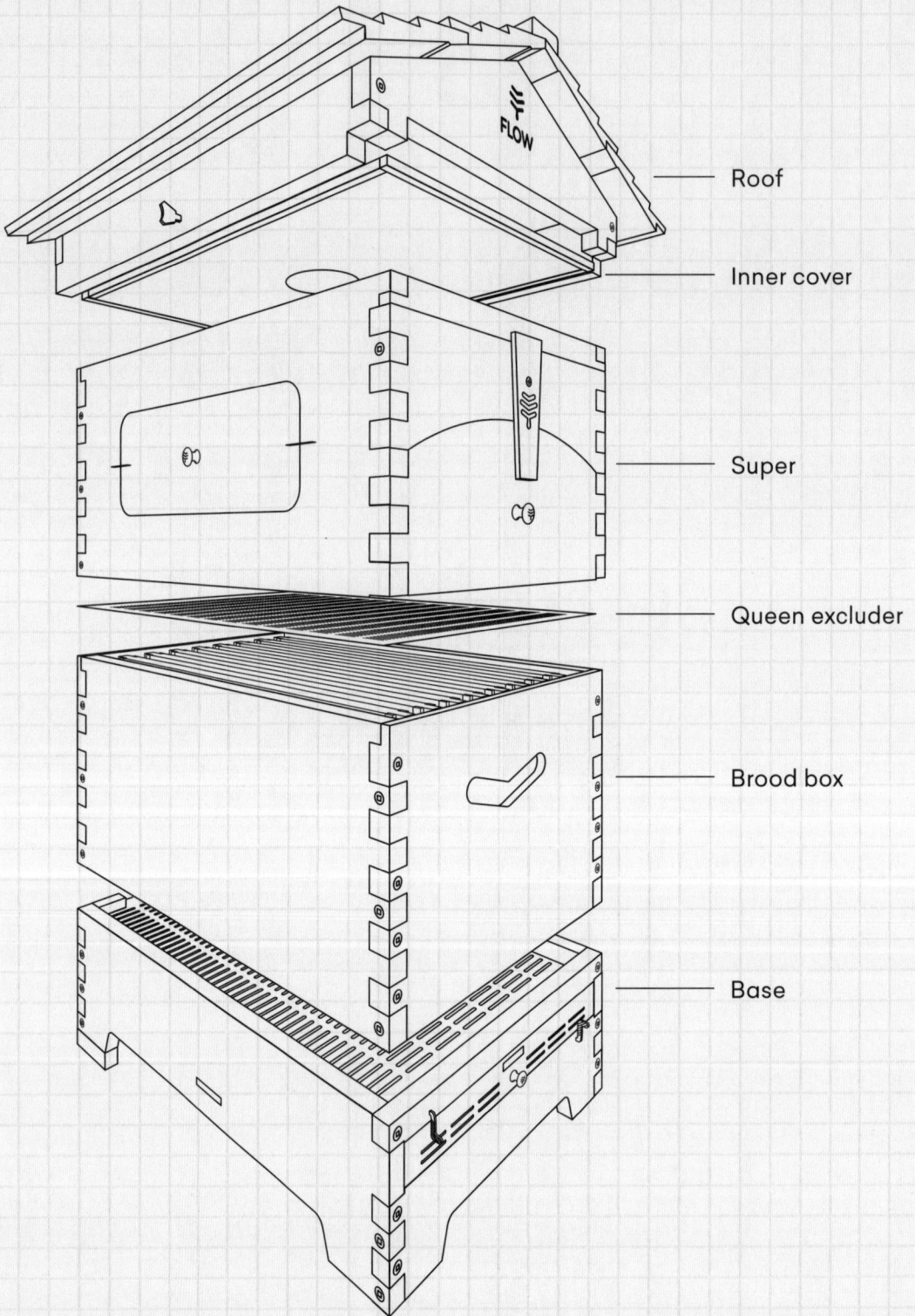

Roof

Inner cover

Super

Queen excluder

Brood box

Base

Honey bees and humans have long been interlinked. It's a win–win situation: we give them care and shelter; they reward us with honey and keep sustaining our food crops. This chapter explains what it takes to set your bees up with a suitable home.

How a modern hive is structured

Many different styles of beehive have been used throughout history. The type most frequently used today is called the Langstroth (or 'Lang' for short). The Flow Hive is a modified version of this style of beehive.

In 1852, Reverend Lorenzo L. Langstroth of Philadelphia patented the design that has become the most widely used style of beehive worldwide today. It features stackable boxes fitted with pre-prepared sets of removable rectangular frames in which the bees build their wax comb. Langstroth's innovation was based on utilising the 'bee space', which is the exact size of the gap that honey bees maintain between their combs, allowing them to move freely and build efficiently.

Langstroth had been inspired by the work of Swiss beekeeper François Huber, who created a hive with combs arranged like the pages of a book. In his 1853 volume *Langstroth's Hive and the Honey-Bee*, the Reverend wrote that he was satisfied with the Huber hive because, with proper precautions taken, the combs could be removed without enraging the bees. Additionally, weaker colonies could be strengthened and those that had lost their queen could have another added. If he thought there was something wrong with the hive, he could quickly work out the problem and apply the proper remedy.

The brood box

The brood box is a bee nursery and larder. It's the heart and engine of the hive, where the queen lays her precious eggs and worker bees are to be found busily feeding larvae, depositing nectar, packing pollen and building comb. The brood box often contains cells filled with honey, usually located towards the edges of the box surrounding the developing young. As well as providing food for the bees, these honey stores act as a thermal mass, helping to keep the temperature steady at around 34–35°C (93–95°F).

The first time you pull a frame from the brood box and see the amazing matrix of cells with thousands of bees going about their work, it really is a memorable moment. And even once you've done it countless times, you'll be amazed how often you see something new and interesting. Checking to see what your bees are up to in the brood box is a central part of beekeeping, and it's a fascinating way to begin understanding the rhythms of the colony. The brood is usually situated in the bottom section of the hive. In Australia, beekeepers typically use a single brood box, but in other regions two brood boxes are often used to allow the colony to rapidly expand if needed and provide space for additional stores to last through long winters or decrease the chance of bees swarming in spring.

Brood frames

Rectangular frames are used to contain the bees' wax comb in a way that allows them to be easily removed from the hive one by one. Without frames, comb would be more likely to be built in irregular patterns that would make it very difficult for the beekeeper to inspect and deal with any issues.

Foundation is a thin sheet of beeswax or plastic imprinted with a hexagonal pattern. It is fixed into a brood frame using wire to guide the bees into making uniformly sized cells (and often to force them to make workers rather than larger drones). Some beekeepers prefer to use foundationless brood frames, which have a small wooden strip at the top from which the bees build their comb downwards. We supply foundationless frames with our Flow Hives, and they can also hold wire.

Top left Bees building natural comb in a foundationless frame. **Lower left** The bees build downward from the comb guide. **Right** Brushing bees off a frame for easier inspection.

I've experimented to see what bees prefer to build comb on. They'll choose empty space first, then wax foundation, then Flow Frames, then plastic foundation.

I definitely prefer using foundationless brood frames because all you need to do is provide the frame and a little wooden strip as a guide and your bees do the rest. With the Flow Hive there's no need for reinforcement wires or plastic foundation as the frames will never be spun in a centrifuge. I no longer need to spend my nights wiring and waxing old frames that have been taking up space in my shed. The frames can simply stay in the hive, and when it comes time to cycle out old comb, it's easy to cut it out and put the frame straight back into the hive. It also makes it super easy to cut out some nice comb from an edge frame for an impressive cheese platter. However, it is a little more work as you should inspect the hive when the bees are starting to build to make sure they're building straight on the comb guides. All in all I would much rather be marvelling at the bees than cleaning, waxing and wiring frames! Allowing the bees to build their own comb naturally in the brood nest is believed to have a health benefit as it lets them size the cells perfectly for their own genetics. And as you're not adding any wax from other hives, it reduces the chances of contaminants being present in the wax, which can happen if foundation is re-used multiple times.

Stu, on the other hand, likes to mix it up and use foundation in a frame or two to give the bees a straight start and avoid cross-combing, which is when they build sideways and connect frames together. (See page 236 for advice on dealing with cross-comb.)

The super

A super, also called a honey super, is – you guessed it – where the bees store honey. In a Flow Hive it's the upper box (or boxes if you wish to use more than one), and it's here that our invention, the Flow Frames, sit. When a new hive is getting established, a super isn't usually added on top of a brood box until the bees have built up their numbers sufficiently. In colder climates it's common to remove some or all of the honey supers during winter so that the bees can maintain their temperature more easily. A queen excluder is a screen that can be used to keep the queen from laying in the super (see page 113).

Varying box heights and widths

In most hives of this style, the lowest brood box is typically a full-depth Langstroth, like the ones we supply as part of the Flow Hive. Above this, box heights can vary. Apart from the full-depth Langstroth boxes, medium and shallow boxes are also commonly used around the world; these weigh less when full, which makes handling them easier. By using box sizes suited to your conditions, you're able to control the amount of space more easily and, if your region has short nectar flows, you can be sure they'll fill and cap the whole frame, which makes extraction of ripe honey easier. We've got information on modifying Flow equipment to suit different widths (for example, UK National) in the Resources section on page 276.

Stu's guide to assembling brood frames

Foundationless frames

Assembling a foundationless frame is fairly straightforward. The main thing is to ensure that it is square, and this can be achieved by hanging the newly constructed frame in a brood box and checking that the sides of the frame are parallel with the walls of the box. For a natural frame, a thin strip of timber is provided to slip into a slot on the underside of the top bar.

Adding plastic or wax foundation sheets

If you decide to use wax foundation in one or more of your frames, do not insert the timber strip. Instead, weave a thin wire horizontally across the frames to attach the wax foundation. The brood frames provided with the Flow brood box have holes in their side members for the wire to thread through. In addition to the wire, you will also need fasteners such as tacks, drawing pins or staples (and a staple gun) to tie off the wire.

The wire usually comes on a spool. Unwind a metre or two, or ask a friend to hold the spool as you pull off what is needed. Thread the wire through the outside of the top hole on one side of the frame and then through the corresponding hole on the other side. Now run the wire down the outside of the frame to the next hole down and pull it through the corresponding hole on the original side.

Again, run the wire down to the next hole, across the frame through the corresponding hole, then finally down to the bottom hole and back across the frame to the bottom hole on the other side. It may have felt a bit clumsy pushing and pulling the wire through all those holes, but now you're ready to tighten the wire.

Tap a tack or drawing pin halfway in next to the hole at the top of the frame that you first inserted the wire through. Fix that end of the wire to the frame by winding it around the fastener you just inserted. Prepare the other end of the wire by adding another tack or drawing pin halfway in next to the hole you've pulled the end of the wire through. Take out the slack so that the wire is fairly taut. Now do a final tightening by pulling on each successive horizontal section of wire. You will get to the final run and hopefully have enough wire left to firmly grip it and pull to get not only the last but all of the wires fully taut. You should be able to pluck them so that they vibrate with a note. Now, holding the tension, wind the end of the wire around the shaft of the bottom tack or pin, and then tap both fasteners fully into place.

There are several ways to attach the wax foundation sheet to the wires:

- press the wire into the wax with a toothed wheel tool;
- use a blade heated on your stove to press the wire into the wax;
- use low voltage electricity (many beekeepers use a car battery) to heat the wires and melt them into the wax.

If you're going to be making more than one or two frames with wax foundation, then it will be worth cutting a piece of plywood or chipboard to rest the foundation sheet on. The rectangular piece should be quite flat and smooth, and sized to easily fit inside your frame. Rest your wax foundation sheet on the plywood so that the wired frame can sit on it. You should be able to slide the gap on the underside of the top rail over the top of the sheet. Now you can use any of the above methods to press the wire into the wax sheet.

Clockwise from top left To add wax foundation sheets to a brood frame, first thread wire through the eyelet; staple the wire to the edge of the frame and tighten the wires; finally, slip the foundation sheet into the frame and press the wires into the wax to support it.

Bees evolved long before humans were building anything. They tended to favour living inside hollows in trees, sometimes in caves or even out in the open on a branch or a cliff face. They'll make their home just about anywhere that's sheltered and suitable for building comb. Today this includes inside the walls of a house.

You might have seen videos of large honey bee colonies being removed from strange places, such as a wardrobe that's been sitting on a verandah or the cabin of a boat. While in the past people would often call in an exterminator if they found that bees had decided on an inconvenient nesting spot, nowadays it's more common for a beekeeper to be called to 'rescue' the colony. A modified vacuum cleaner is often used to get the colony into a box, ready to be moved into a hive in a new location.

Harvesting honey from an urban hive.

Painting your hive

If your hive is made from a timber other than Western red cedar, it's almost definitely going to be necessary to paint it with a sealant. Use at least two coats of good-quality exterior paint for the most effective protection from the elements. Make sure you don't forget to paint the top and underside of your hive's roof, no matter which timber it's made from.

Beekeepers who like a natural timber look often use an oil finish on their Western red cedar hives. In wet climates, oiled timber can develop mildew. This won't affect the structure's integrity or the bees, but not everybody likes how it looks. If you're after a natural finish without mildew, ask your local paint supplier for a long-lasting outdoor sealant or timber decking product.

Many people prefer to leave the inside of their hive untreated, especially if the finish they're using contains strong synthetic chemicals. The bees will often coat surfaces inside the hive with wax in any case. Before installing your bees, the finish should be left until it's completely dry and any strong smells have dissipated.

Positioning your hive

You'd be surprised at the variety of places that people manage to keep bees! One of the first things you must consider as a new beekeeper is where to put your hive. Wherever you place it, you'll need to follow some simple guidelines:

- Have room around the hive to be able to inspect and harvest easily.
- Make sure the bees' flight path isn't going to cause problems (see more on this on page 94).
- Consider kids and pets on the property, plus your neighbours and any local regulations.
- Keep your hive level in the side-to-side direction (more on page 95).
- Raise it off the ground away from ground predators, and to a height that's comfortable for you to do your brood inspections.
- Make sure the hive is protected from strong winds or is strapped down.
- There should be a source of fresh water nearby.

When I was growing up we had beehives on the roof of our carport in Canberra. My brothers and I fixed a ladder to the wall so that we could easily access the hives. I loved seeing the amazement on the neighbours' faces as we climbed up there in our homemade bee suits.

— Stuart

Sun or shade?

Bees can deal with a fairly wide range of temperatures, but if you have a choice between full sun and full shade, sun is often better as it helps to reduce dampness and keep any pathogens, such as chalkbrood, away. However, if you live in a very hot climate, then afternoon shade in summer will likely be welcome. Commercial beekeepers often like to position their hives so the morning sun shines in and gets the bees up working earlier. In a backyard, although an easterly aspect is nice, other factors may be more important in terms of which way your hive faces.

Coming in for landing

Consider your bees' daily flight path. Even if your bees are generally docile, it's best if they're not getting tangled in your hair when they're on the way back from a foraging trip. Make sure your hive is located away from busy areas and avoid placing it with the entrance facing a walkway. Ideally, you'll set your bees up with a couple of metres of clear space in front of their entrance to make it easier for them to take off and land. However, their flying skills are incredible and they can make do with a small amount of space if they have to. In situations where there's not much space, some backyard beekeepers position the entrance of their hives near a fence to make the bees fly upwards and above the heads of passing people.

Have you ever noticed or wondered about those little yellow dots that you sometimes find on your car, especially at certain times of the year when you park under a flowering tree? If you scratch and sniff, you will soon work it out... it's bee poop. (Maybe don't tell your neighbours about this!) Because your bees are very hygienic and won't poop till they fly out of the hive, it's best not to point the entrance of your beehive towards cars, washing lines or anything that you don't want the bees to 'decorate'.

Forager bees landing at a hive. Ideally, set up your hive with a couple of metres of clear space in front to make it easier for bees to arrive and depart.

Another thing to be aware of is that bees are attracted to light, so keep the hive away from strong outdoor lamps or they might miss out on their bee-auty sleep. (Hey, it's a bee book … we tried to resist, but there's going to bee a pun or two!)

Are there bee predators I need to watch out for?

You'd be forgiven for thinking that bees' stingers would make them unappealing as potential snacks for other animals, but this is not always the case! The notorious cane toads like to visit our hives in the wee hours to gobble up any guard bees who are on night shift at the hive entrance. For this reason, keeping the entrance at least 300 millimetres (1 foot) off the ground is a good idea. In North America, some beekeepers have trouble with skunks. If this is an issue for you, you might want to raise your hive even higher.

Should I fence my hive?

While a sting on the nose will quickly teach your beloved fluffball not to approach the hive, some pet owners choose to fence off their hive to keep animals on the property safer. Everyone knows that bears love honey. In some parts of the USA and Canada, electric fences are used to keep bears from raiding hives. Livestock have also been known to let their curiosity get the better of them and can end up knocking over beehives when they get too close. If this happens and the hive is not put back in its correct position within a day or so, the colony is likely to abscond. A fence can be helpful to prevent this scenario. Make sure no pets or farm animals are ever tied up close to a hive.

Keeping it all on the level

It's important that your hive is level in the side-to-side direction so that when the bees build their natural brood comb downwards it will hang straight inside the foundationless brood frames. In the front-to-back direction, your Flow Hive should have a 3-degree slope, preferably towards the harvesting end (the opposite end to the bees' entrance), to allow the honey to flow out smoothly during harvests. Our baseboard has this slope built in, so that when the baseboard is level the hive will be sitting at the correct 3-degree angle.

Beekeeping equipment

Bee suit

Back in the day, we used to wear homemade protective gear such as washing up gloves with rubber bands around the wrists, and veils sewn onto old hats! It was effective enough, but nowadays good quality safety gear is more easily available. Modern bee suits include gloves, a mesh veil or hood and a jacket or full-body jumpsuit made from sting-resistant materials.

If you're new to beekeeping, it's best to always protect yourself with a good bee suit and gloves while tending to your bees. Once you're familiar with their temperament and behaviour, you may feel comfortable being close to your hive without wearing protection. However, whenever you open a hive (or mow the grass close by), we always recommend a suit and gloves to minimise the chances of being stung. Make sure there's nowhere that bees can get in, so always check the areas around your ankles and where the gloves meet the suit. Make sure when you zip up your suit that the zippers overlap and have velcro covering them. To have an enjoyable experience you want to stay nice and calm, and that can be difficult if there are bees crawling up your pants! Many of our customers also like to keep spare veils or suits available for visitors who are curious to take a look at their hive.

In the context of thousands of years of honey gathering, the bee suit is a relatively new invention. However, beekeepers have been using some form of protection from stings for hundreds of years. Some of the best records of early bee suits can be found in books and artworks. The drawing *The Beekeepers and the Birdnester* (see page 59) shows three apiarists wearing long tunics and hoods with attached wickerwork masks to protect themselves from bees, which was a common (if not 100 per cent effective) safety precaution at the time.

The smoker

A smoker is an essential part of your beekeeping toolkit as it helps calm the bees and causes them to move out of the way. Smoke interferes with their sense of smell, masking the alarm pheromone given off whenever they sense danger. Without smoke, opening the hive may send the worker bees into defence/attack mode. The impact of smoke usually lasts around ten to twenty minutes.

The smoker contains a small fire chamber, a tight-fitting lid with a spout, a bellows and usually a heat guard. It's essential that your smoke isn't too hot and has no toxic fumes, so avoid burning anything in it that contains plastic, heavy oils, hair, paper or feathers. Examples of appropriate types of dry fuel are:

- dry leaves
- clean hessian
- bark or pine needles,
- wood shavings
- dry lawn clippings
- rolled up clean non-printed cardboard
- commercial fuels made from pulped paper, wood pellets or compressed cotton.

Beekeepers have used smoke for honey harvesting since ancient times, carrying smouldering sticks with them as they climbed trees to raid wild hives. Native Americans figured out that they could burn puffball fungus to pacify honey bees when gathering honey, and this practice later inspired some (gruesome) experimentation to see if the smoke could be used as a general anaesthetic on animals. Up until quite recently, European beekeepers had holes in the masks of their bee suits through which they could insert a special metal pipe and use it to smoke herbs such as tansy to calm the bees. The type of smoker that is most frequently used in modern beekeeping, featuring a bellows attached to a tin burning chamber, was invented in 1873 by an American Quaker named Moses Quinby.

There's an art to lighting a smoker, and every beekeeper develops their own favourite technique. One way that works well is to loosely pack in the fuel, light it and gently puff the bellows so you get nice flames. At this point, it's tempting to jam a whole lot more fuel in, but that will usually put out the fire and you'll have to start over. It's more reliable to add a little more loosely packed fuel and continue puffing. Repeat this process till you have built up some nice embers. Then you can go ahead and pack the fuel more tightly, and your smoker should keep going, producing nice bellows of smoke. Once it's properly lit, close the lid. This should put out any flames. Puff the bellows vigorously to keep the fuel smoking properly and check that no flames or sparks are coming from the spout. If they are, simply cover the spout for a few seconds (with something that won't burn, obviously!) to put out the flames.

Thick smoke that is cool to the touch is best. You can smoke your hand to test the temperature, which will also help to mask your scent. Once the smoker is smoking properly, you can pack in plenty more fuel while continuously pumping the bellows, giving the occasional puff after that so it stays lit. Depending on how long your beekeeping session lasts, you might have to top up the fuel once or twice as you go.

Always be careful handling your smoker when it's active as it can get hot. Hold it by the bellows, preferably wearing gloves, and use a tool to open the lid to avoid burning yourself. Keep water on hand and avoid using the smoker in high fire danger conditions. When you've finished, it's crucial to ensure that the fire is out entirely and any ashes or embers have been removed before the smoker is put away.

Hive tool

A hive tool is a piece of metal that beekeepers use to make it easier to manipulate the hive and its components. One of the most common designs has a J-shaped hook at one end and a chisel-like tip at the other. It's helpful for many beekeeping jobs, from lifting frames out of the super or brood box to prying apart parts of the hive that the bees have glued together with propolis (or bee glue; see page 180). A hive tool is also useful for removing burr comb (comb that has been formed at the edges of the frame). Flow Frames are designed to be easily lifted using the common J-style hive tool.

Clockwise from above Check the smoker's temperature: it should be cool to the touch; the J-style hive tool for lifting, prying and scraping; the Flow Super Lifter makes opening a beehive easier.

Getting and installing bees

If you're new to beekeeping, unless you've done a course or helped a beekeeping friend in the past, the day you install your bees will be your first time getting up close and personal with several thousand furry flying creatures all at once. Exciting! Remember to be aware of their sensitive nature and the basics of safety, and to enjoy this moment of getting acquainted with your new winged friends.

Some people mistakenly think all they need to do is place a hive in the garden and bees will turn up. If that's your strategy, you could be waiting a very long time ... If you would prefer to have some bees in your beehive, the easiest way to procure some is to source them from a reputable local supplier, either in a 'nuc' or as a 'bee package'.

Nucs

In beekeeping, 'nuc' is short for 'nucleus'. This is a small established colony in a box that comes complete with frames, brood, bees, a queen, pollen and honey stores. When you get your nuc, transfer the frames into the middle of your hive's brood box, carefully placing the nuc frames close together. Keep them in the same order they arrived, as the bees typically like to keep the developing brood at the centre and food stores more peripheral. You then add extra empty brood frames on either side of the nuc. It's important that all the frames are pressed tightly together and centred in the box so that the extra space is on the edges. As your nuc develops, the bees will build comb in the empty frames and continue to expand their brood nest. Always fill the box with frames as your bees will be quick to fill any spare space with random comb that you can't lift out for your inspections.

Bee packages

A 'package' of bees includes a mated queen and around 10,000 bees in a ventilated box. Amazingly, in many regions you can order these in the post. (You may get some odd looks from the delivery person as they deliver a buzzing box of bees.) The queen and a few of her attendants are often separated from the other bees in a small cage with a plug of candy blocking the opening. Make sure your brood frames are in place – packed together towards the centre of the box – and place the queen cage horizontally between the upper bars of two frames. Shake the other bees into the brood box and put the roof on the hive. Over the next day or so the bees will chew away at the candy, releasing the queen and her attendants to join the rest of the colony. Don't open the hive for around five days or a week; disturbing them could decrease the likelihood of the queen being accepted. The colony should start foraging within a day – keep an eye out for pollen on the legs of bees returning to the entrance. By the time you open the hive to do your first brood inspection there should be eggs in the brood frames, although you might need bright sunshine or a flashlight to be able to see them. See page 139 for more details.

Splits

Splitting is a great way to get your first hive started. In the springtime bees really build up in numbers, and hives are likely to swarm if not given more space. By taking a split from a beekeeping friend you will often be helping them by making more room in their hive and limiting the chance of their bees swarming.

A 'split' should come from a colony that's large and healthy, with frames of capped brood to help the colony grow quickly. The best time to split a hive is when there are plenty of bees in the hive and lots of forage around. In most regions this means spring, but in some places you can take splits at other times of year too. It can be a good idea to order a queen from a queen breeder for your new split. This way you are much more likely to get nice, calm productive bees. Alternatively, you can allow the bees to raise their own queen by giving them a frame that has some bee eggs in the cells. For more on splits see page 202.

Leaving the hive

Swarming is how a beehive reproduces itself. In spring or summer, if plenty of nectar and pollen are coming in and the hive is crowded, the workers are likely to raise at least one new queen by building a specially shaped cell and feeding the larva in it with royal jelly. After her royal highness has matured, roughly half the bees in the hive will fill their bellies with honey and then depart. They fly in what looks like a cloud and settle somewhere, usually on a tree branch, with the old queen safe at the centre.

Now it's time for scout bees to find a suitable new home. They search all over, evaluating potential sites based on their size, location and safety. When a scout finds a good spot, she returns to the swarm and performs a 'waggle dance' (see page 46) to communicate the location she's chosen. Others will follow her directions and investigate her proposal for themselves. If they like it, they'll go back to the swarm and perform the same dance. Several dances might be taking place on the cluster of bees at one time, and as soon as one of the dances gains enough 'followers', the swarm will fly directly to the new spot (literally, 'make a beeline' for it) and begin building comb.

Professor Thomas Seeley, who first described this process, says that the method of choosing a new home demonstrates sophisticated forms of consensus decision-making including 'collective fact-finding, vigorous debate, consensus building, quorum and a complex stop signal enabling cross-inhibition, which prevents an impasse being reached'. If this is a measure of democracy, bees are probably better at doing politics than humans most of the time.

Catching a swarm

If you spot a swarm resting in a tree or a bush, it's a great opportunity to provide them with the perfect home. And swarm catching is definitely one of the most exciting parts of beekeeping! Although a large cluster of bees can look intimidating, they're usually quite docile at this time. But not always! So you'll still need to suit up and wear gloves, and also be careful if you're working at height. Trim any branches that are in the way, taking care not to disturb the bees. Next, place a brood box directly underneath the swarm and give the branch one good, sharp downward shake so that most of the bees fall into the box. You might need to give the swarm a couple more shakes to get the majority of the bees into the box. Carefully add brood frames, keeping them packed closely together and centred in the box so that the bees will build their comb correctly. Then put the roof on and make sure the box is fairly level. Assuming the queen is in there, any stray workers will begin marching in. If you can, remove the branches they'd been sitting on so that the pheromones left there won't tempt them to return to that spot. You can then come back and move the bees to a suitable location as soon as they are all in the box, or you can leave them there and move them at a later date. For details of how to move a hive, see page 253.

Sometimes it's ridiculously hard to catch a swarm; other times they'll just land right in your garden where you can box them very easily. I once chased a swarm of bees through the forest and they landed high up in a tree. I climbed the tree and cut the branch down, but when it hit the ground the bees went everywhere, of course! I had to repeat the process five times before eventually lowering them gently into a box. For years afterwards, eating honey from that hive always reminded me of that day in the forest.

—Cedar

Bait hives

People often ask: 'If I put a hive out in my garden will bees move in?' The answer is, it's unlikely, but you can do a few things to increase the possibility of this happening by creating what we call a bait hive. When a swarm of bees is voting on a new home, size and location are key to their decision. To increase your chances of success, place a single brood box complete with brood frames, lid and baseboard 2 to 4 metres (6 to 12 feet) off the ground near existing beehives. You can also add a little lemongrass oil which works as an attractant. If it's springtime and there are lots of swarms, you may be lucky.

Now that you know the basics of how a beehive works and how to install your bees, let's take a deep dive into the Flow Hive.

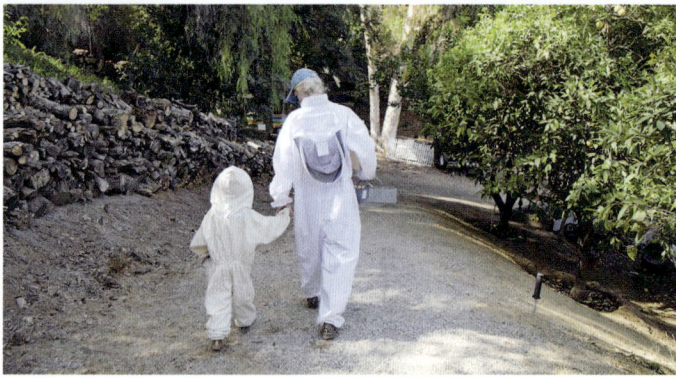

'We lean on each other for confidence – it's been great having a friend learning with you and growing with you and helping you.'

Beekeeping with friends

Brooke, Stephanie and Lemon,
California, USA

Brooke started beekeeping at the same time as her good friend Stephanie, and they've loved supporting each other on the journey. Brooke's daughter Lemon is a budding beekeeper who thinks every bee is a queen. To see more of their story, check out the Resources section on page 276.

'My whole goal is to teach my girls to be sustainable and to be able to feed themselves and how to grow food and cook food in a healthy, sustainable way.'

Scan here for videos relating to this chapter

4

About
the
Flow Hive

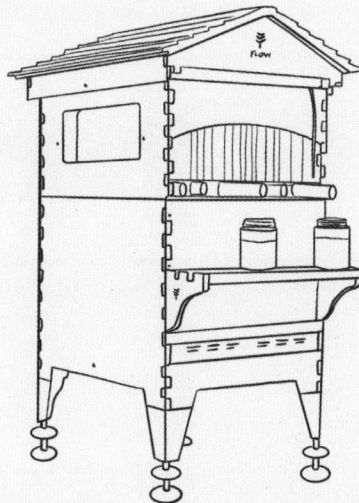

The Flow Hive was born from a simple idea: can we easily and gently harvest the honey directly from the hive while the bees are left to bee, and skip the need for a conventional extracting room and all the palaver and heavy lifting that comes with it?

It's been an amazing journey to achieve that idea, and even more amazing to share this new way of harvesting honey all around the globe. It's still mesmerising every time we see the golden liquid flow right out of the hive.

One of the many unexpected things we've found along the way is how the taste and colour of the honey can vary so widely from one Flow Frame to another. This is because the bees usually fill each frame sequentially with nectar from whatever is in bloom at the time. One of the many joys of beekeeping with a Flow Hive is being able to isolate and taste all the different flavours of the season.

Building and finishing your Flow Hive

When you get your Flow Hive, all the components and some handy tools are included in the box. A drill and a set of clamps can be helpful but aren't strictly necessary. We know how frustrating assembling flat-pack furniture can be, so we've created detailed assembly videos for every version of the Flow Hive. Go to the Resources section on page 276 for links to these videos.

As mentioned on page 93, you must treat or paint at least the roof. This is best done after the basic structure has been assembled but before all the fittings are attached. Once the paint is completely dry you can attach knobs and windowpanes.

Adding Flow Frames to an existing hive

If you're a beekeeper who's already using Langstroth boxes, you don't have to buy a whole new hive, as you can simply add a Flow Super to your existing hive or modify a super and add Flow Frames to it. Note that you'll need to consider the slope. For optimal draining during harvesting, the hive should tilt slightly backwards at a slope of 2.5 to 4 degrees. You can either chock up the front of the hive when it's time to harvest or, better still, have it sitting on a permanent slope. However, in that case it's important to make sure that water entering the hive at the front doesn't become an issue. Using a sloped landing board and screen bottom board can minimise the chances of this.

We love to see how creative beekeepers can be when finding new ways to use our invention. There are links to easy-to-follow instructions on modifying Langstroth hives in the Resources section on page 276.

The Flow Frames

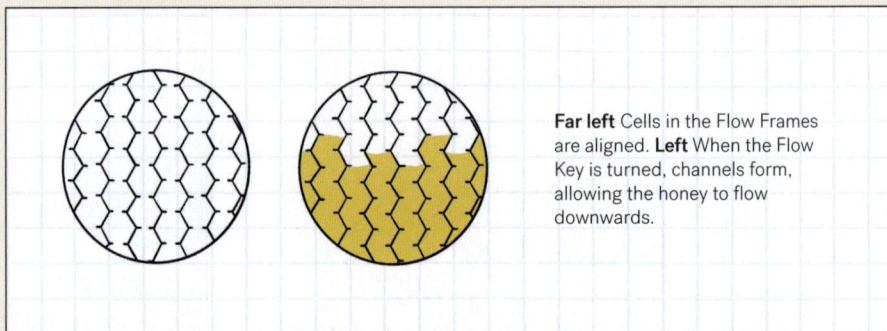

Far left Cells in the Flow Frames are aligned. **Left** When the Flow Key is turned, channels form, allowing the honey to flow downwards.

Our Flow Hive invention is where the magic happens! The Flow Frames allow you to tap honey directly from the hive without needing to take out the frames for processing. This is a big change from conventional harvesting methods.

Flow Frames provide the bees with a partially built honeycomb matrix. The bees complete the partly formed cells with their wax, then fill them with nectar and cap the cells once the nectar has been turned into ripe honey. When the Flow Key is inserted and turned 90 degrees, the cells split and the honey flows downwards into a collecting trough then out through a tube into a jar. Afterwards, the bees will chew the wax cappings off, recycle the wax and start refilling the cells with nectar.

We considered every detail of the Flow Frames carefully and gave a lot of thought to the design in order to achieve our goal of easy harvesting directly from the hive and a gentle process for the bees; for example, we built in a gap between the moving parts which the bees bridge with their wax in order to complete the cells. This ensures that if there happen to be bees in some cells during harvesting, they're unlikely to get a leg or a wing caught when the cell parts realign.

Another thing is that the head of the moving blades is made to exert maximum force on the frame when the key is turned, though if honey has crystallised in the frame, the plastic will flex rather than break.

Cell sizing

You may notice that the cells in the Flow Frames are bigger than the worker brood cells or typical foundation-sized cells of 5.3 millimetres (⅛ inch). When building comb away from the brood nest for honey storage, bees will usually build cells that are larger and deeper for more efficient storage. After lots of counting and measuring cells with callipers, we chose a cell size close to 6 millimetres (¼ inch) to mimic what the bees make naturally for honey storage away from the brood nest. This size makes it less enticing for a queen to lay eggs in the frames, because the cells are too big for worker bee eggs and a little on the small size for drones. However, we have found some queens will still lay eggs here, so we recommend using the queen excluder.

As with all their other tasks, when honey bees construct their comb they work together. A worker will gorge on honey, then a gland converts the sugar from the honey into a waxy liquid, which is secreted on her abdomen and solidifies in the air. Other bees take these tiny flakes of wax and soften them in their mandibles before carefully adding them to the comb. Beeswax is expensive for the bees to create: any amount of wax they produce requires them to eat around seven times as much honey.

While building comb, the bees will often form a chain by linking their legs. This behaviour is called 'festooning', and to this day nobody knows exactly why they do this. Some experts think it helps with measuring distances or acts as a useful scaffold.

Although comb is usually composed of hexagonal cells, bees will just as happily build four-, five- or seven-sided cells whenever those shapes are more suitable. Hexagons are the most efficient possible structure in terms of storage space relative to strength and material, but the bees don't measure a 120-degree angle at each corner. They start by building a circular shape and then heat the wax up so that it melts enough so the cell junctions 'flow' and become perfect hexagonal corners.

In our experimentation, we've found that rather than being 'hardwired' to build perfect hexagons, bees will build the most efficient shape under the circumstances. For instance, if you give them a grid to start with, they'll fill it to make square cells. We've even seen them make downwards-facing square cells that somehow hold honey.

Experiments by renowned Swiss beekeeper François Huber in the 18th century showed that bees will quickly adapt to obstructions placed in their way, readily altering their design to work around these challenges. And they do all of this in the dark!

Leak back point

When you look at the bottom of the rear end of a Flow Frame, where the tube for harvesting is inserted, you'll see a gap. There's a reason for this: we left a space so the last few drops of honey from the harvest can leak back into the hive. It's fun to watch the bees use their long feathery tongues to clean up the last drops of honey after a harvest. The harvesting tube has a little tongue on it that unblocks the leak back point if the bees have plugged it up. Of course bees will be bees and wax up everything over time, so you may need to manually clean out this leak back point occasionally to stop honey building up in the trough area. This can be easily achieved by inserting a harvesting tube or any pointy object.

For information on how to maintain, clean or fix your Flow Frames, see page 248.

Features and accessories

Here's an overview of some important features of the Flow Hive.

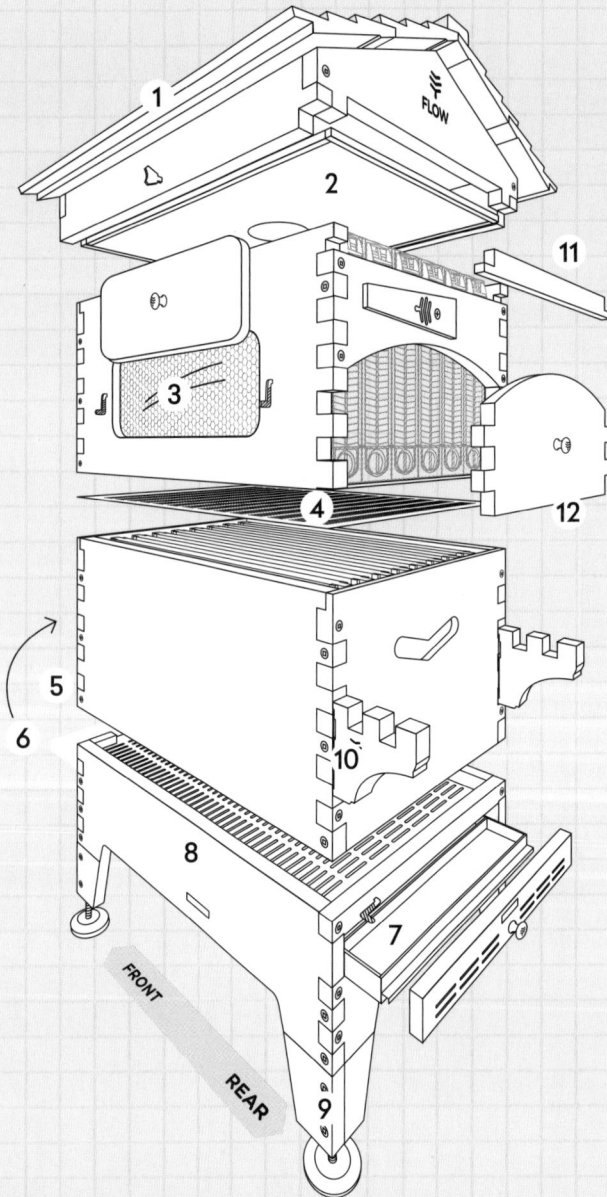

1 Gabled roof

The gabled roof adds another layer of protection from the elements over the inner cover, helping your bees to keep a stable temperature. When you assemble the roof, be sure not to leave any gaps and then give it two good coats of exterior house paint. Some people like to apply a waterproof caulk to the joints for an extra tight seal.

2 Inner cover

The inner cover sits between the roof and the top box. A hole in the inner cover allows you to easily feed your bees during lean times and can be easily closed off when not in use (otherwise the bees will build comb in the roof cavity).

3 Observation windows

The observation windows in the Flow Super have proven to be a great feature. By observing your bees filling the cells and capping them with wax, you can get a good idea of when it's time to harvest. You can also see when they're hungry and consuming their stores. Getting a window into their world is amazingly helpful in gauging how well the colony is doing.

4 Queen excluder

A queen excluder is a barrier used to prevent the queen from entering the super from the brood box below. It's sized with approximately 4.3 millimetre (⅙ inch) gaps, which worker bees can move through but queens and drones can't. This keeps your queen from laying eggs in the Flow Frames. Most Langstroth-style hives utilise a queen excluder; however, some Flow Hive beekeepers choose not to because many queens will never lay in Flow Frames anyway. If you choose not to use an excluder you'll need to do routine inspections to make sure there's no brood in the super.

5 Entrance reducer

An entrance reducer minimises the entry to a hive to protect it from pests, predators and bees from other colonies looking for nectar and honey (these are called robber bees). It also reduces drafts in colder climates and can be used to fully close up the entrance should you need to move your hive.

6 Weather guard

A weather guard (pictured on page 114) can be useful to protect the hive entrance and keep the landing board safe from the elements. If your region experiences a lot of inclement weather, your bees will appreciate the extra cover!

7 Multifunctional tray and ventilation control

The tray is useful for making observations and assists with pest management by allowing you to easily add traps or treatments. The tray cover is also a ventilation control. Inserting it one way up or the other allows you to increase or decrease the airflow from beneath, depending on external weather conditions. You can also remove the tray altogether for maximum ventilation.

8 Sloped base, mesh screen and adjustable hive stand

For the honey to flow out smoothly when you're harvesting, the hive should sit at a 3-degree angle front to back. The base is designed so that when it's level the hive is tilted correctly for harvesting. The mesh screen in the base is helpful for trapping pests.

A stable non-wobbly hive is essential. Our adjustable stand makes this easy and keeps the hive off the ground so it stays dry and out of reach of most ground-dwelling predators. Each leg can be set to the exact height required, and inbuilt spirit levels at the side and rear of the base make it easy to find your level (which is doubly important when using foundationless brood frames so that the bees can build their comb straight down).

9 Ant guards

Ants in a hive are not necessarily a problem for the bees, but in some regions they can be and you might prefer to keep them out. Our ant guards create a barrier between the ground and the hive. You can fill these with cooking oil or Vaseline. You will also need to remove any foliage that touches the hive and sweep away all the ants and eggs from all areas (don't forget under the roof). Once you have done this a few times your ant issue will go away (until next time).

10 Harvesting shelf brackets

A shelf is convenient for resting your jars on while harvesting. By attaching brackets to the back of the hive, you can use the cover of the rear observation window as a nifty harvesting shelf.

11 Flow Key access cover

This is removed when it's time to harvest so that the Flow Key can be inserted into the top of the Flow Frames.

12 Rear window cover

This can be removed to help gauge how much honey has been stored and to allow harvesting tubes to be inserted in the bottom of the Flow Frames. It can also be used with the harvesting brackets to create a shelf for your jars.

A queen excluder on top of a brood box.

A Flow Entrance Reducer.

Flow Harvesting Shelf Brackets in use with the end cover.

Flow Ant Guards.

A Flow Weather Guard.

Flow Inspection Frame Rests hold brood frames while you work.

I grew up with hives that had a solid floor with the base tilted so that the entrance was lowest. This allowed any rain to dribble back out and made it easier for the bees to haul out any detritus. Many beekeepers are used to this way of setting up the hive and find Flow's 'reverse tilt' a bit confusing. But as the mesh floor allows rain and detritus to drop into the tray below, it is no problem to tilt the hive 'backwards' so the Flow Frames drain properly.

—Stu

Now that we've covered the important specifics of the Flow Hive, let's move on to the next topic, which is possibly the most important chapter in the whole book: how to best care for your bees.

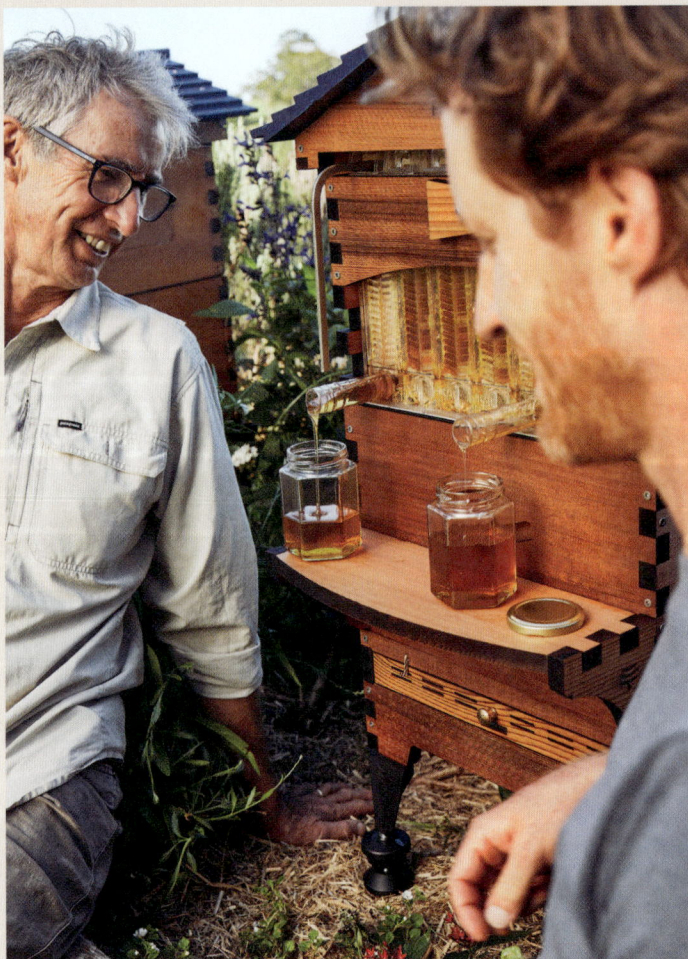

Right Stu and Cedar harvesting honey.
Above The bountiful harvest.

MEET THE BEEKEEPER

Beekeeping together

Paras, Rajeshree and Aadi, Essex, UK

For Paras, Rajeshree and Aadi, beekeeping is a fun family experience. They've enjoyed learning from friends at their beekeeping club, but most of all learning from the bees themselves.

To see more of their story, check out the Resources section on page 276.

'Having bees as a family, for us it's like a little project where we've all got our roles. And getting our son Aadi involved – I want him to fully embrace Mother Nature.'

Scan here for videos relating to this chapter

5

Caring for your bees

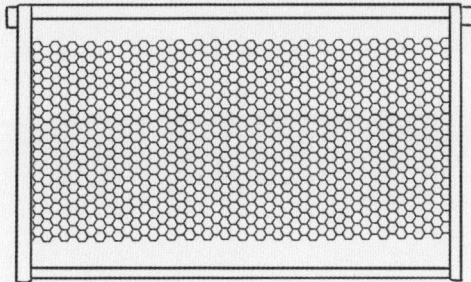

One of the extraordinary things about beekeeping is, no matter if you're a 'newbee' or a seasoned expert, whenever you open a hive and delve into the inner workings of the superorganism, there's a great chance you'll see something new. You never get bored watching the bees at work.

Like caring for any living creature, looking after bees is both a science and an art. Getting to know your new non-human friends is the first step towards understanding their needs. But be warned – spending time with bees and getting drawn into their world can be a bit addictive!

Watching and learning

You can learn a lot about your bees by observing what they're doing from outside the hive. It's a great thing to do while enjoying a cuppa! While watching the bees flying off and coming back, you can't help but be reminded how connected everything in nature is. As you get into the habit of checking in on them, you'll soon become familiar with their habits and notice little changes as they arise.

One of the first things you might notice is foragers returning to the hive entrance with colourful pollen balls on their hind legs. That's a sign of good local forage. The colour of the pollen will depend on what's in flower nearby. Do you know which trees and flowers are blooming in your area right now? It's fun to find out and a great way to get more in tune with your environment as time goes on.

On fine days during spring and summer, your bees should usually be quite active around the hive. If you have more than one hive, compare their activity levels: each colony is unique and you might notice things like one hive being more lively than others at different times of day.

Orientation flights

Young bees need to get to know the local area, so when they come of foraging age they begin embarking on orientation flights, most often during the late morning or early afternoon. You'll see them flying in front of the hive in a circular or figure-eight pattern. They gradually increase their distance from the hive while maintaining visual contact with it – this allows them to get to know the location and landmarks around their home. They'll also perform short flights in various directions around the hive, exploring the area and memorising visual cues. Seeing your bees on their orientation flights is a sign of a healthy colony, as it means that young bees are successfully completing their development and are getting ready to take on foraging and other tasks outside the hive. A burst of orientation flights is sometimes mistaken for bees preparing to swarm, but it's normal to see quite a lot of bees flying near the entrance, especially when the sun first comes out after rain. A healthy colony will have bees coming and going all day long, unless the weather is very cold or rainy.

Entrance behaviours

You may see your bees fanning their wings at the hive's entrance. This not only helps to ventilate the hive, but also helps to dry the collected nectar into thick, sweet honey. You might also spot them fanning their wings while tilting the tip of their abdomen downwards to expose a scent gland. These glands are spreading Nasanov pheromone, which helps guide foragers back to the hive.

When bees gather outside the hive in a cluster, either spread evenly across the surface or more densely clustered in a ball, it's called 'bearding'. This helps keep the temperature inside the hive cool on a hot day by making room for adequate ventilation, and can also occur if the number of bees in the hive is getting quite high. Bearding is normal during hot weather, and when it's really hot and crowded they might cover the entire front of the hive. While this is a sign of a strong, healthy hive, make sure you check for overcrowding. If you look in the observation windows and can't see the comb because there are so many bees, it might be time to take a split or make some room by adding another box to prevent swarming. See page 202 for instructions on splitting a hive.

Once in a while you might notice worker bees gathering in rows near the hive entrance and moving rhythmically back and forth, using their front legs and mandibles to scrape against the wood. This is called 'washboarding', as it can leave horizontal lines in the wood that make it look a little like an old-fashioned wash-board. It's more commonly seen in times of nectar dearth, and usually happens in the evening when there's less foraging happening. Nobody really knows why bees do this, but some people theorise that they're cleaning. Others think it's a form of communication or that they're depositing scent on the hive. Maybe somebody will figure it out someday or it will remain just another fascinating bee mystery!

Right Watching the hive entrance can give you clues about the health of your hive. **Opposite page** Bees 'bearding' on the front of a hive may be because of heat or overcrowding. **Following pages** A bee dispersing Nasanov pheromone; a worker carries a bounty of white pollen.

Checking the multifunctional tray and around the hive

With a Flow Hive, you can check the tray for debris. Look out for mites, beetles or larvae. Seeing many of these could indicate you need to check inside the hive for pests and take appropriate action. In many countries the tray is used to count mite drop and ascertain whether a treatment is needed (see more on mite monitoring on pages 241, 243–244). Little white flakes of wax are a good sign, indicating that new comb is being built. Little circles of silk cocoons are a sign that plenty of new bees are chewing their way out of their cells.

When you put the tray back in, check to see if any bees are on the underside of the screen: you don't want them caught in the tray area. Also make sure the vented cover is pushed in all the way and secured, as this creates a seal that stops bees getting into the tray area. Early morning is the best time to check the tray, as the bees aren't as active then, so they're less likely to get in there.

Don't be alarmed if you see a few dead bees in front of or near the hive. It's natural for a certain number of them to expire inside the hive each day. Workers may only live for a month or so, and when they die their bodies start giving off fewer pheromones than before. Undertaker bees find them in the dark in as little as half an hour and will remove them from the hive before they start to pose a health hazard. When you do see dead bees, take a closer look. Are they primarily workers or drones? If there are a lot of dead drones, it may indicate the colony is stressed, but more often signals a change of season: the colony can sacrifice drones in times of dearth to make the larder last longer. If you see a lot of dead workers piled up high and extending out from the hive like a carpet, it's a sign of something more serious, such as exposure to pesticides (see page 230 for advice on how to identify and deal with this issue).

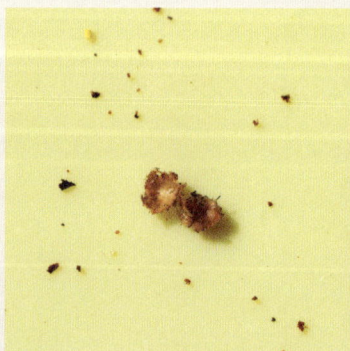

Right The Flow Hive's multifunctional tray. **Above** Wax cappings in the tray indicate that young bees are emerging from their cells. **Opposite page** Cedar examines a brood frame. **Following pages** Stages of brood development in the cells, from left: eggs, uncapped larvae and capped worker brood.

Brood inspections, step by step

The brood box is almost always the bottom box in a Flow Hive and in most Langstroth-style hives. As we mentioned in previous chapters, the brood box (or boxes) is where the bees raise their young. **Inspecting the brood is the most important activity in beekeeping**, and it's when you get to see the wonder of the superorganism up close.

As their keeper, it's your role to support your bees as best you can. There are some key things that you'll want to check on regularly to ensure your colony is well: primarily to make sure the queen is present and laying, and that there are no signs of pests or disease. When and how often to inspect will depend on local conditions and the time of year. Local beekeepers will typically be able to advise you on best practices for your location.

The perfect time to inspect a hive is midmorning to midafternoon on a warm, dry, still day. It's best to avoid opening the hive if it's extremely windy or raining. Bees don't like windy weather or getting wet as, among other things, it makes flying difficult. When all the foragers are inside out of the rain and wind with nothing to do, a normally placid hive is more likely to get a bit grumpy. Maybe humans are kind of similar to bees in this way ...

If it's your first ever brood inspection, it's nice to have an experienced beekeeper by your side if possible. With your first inspection, you don't have to try to learn everything all at once – just get comfortable opening a hive and taking out frames, and then putting them back in. Once you've got the first inspection under your belt and built up some confidence, you can aim to check the frames next time and start looking for specific things.

Try not to squash any bees as you go, though don't be too upset if you do – it's something that happens in beekeeping. Ultimately what you're doing is aimed at keeping the colony healthy, so it's all for the greater good.

1   Gearing up to approach the hive

Visiting your bees is always an opportunity to take a moment out of the busy-ness of life, learn something new, marvel at the intricacy of how the superorganism functions, or do all of the above at once. Some beekeepers say that the bees will pick up on your mood. We like to be calm, work slowly and gently, and enjoy being present with the bees.

Stu once visited a school in the UK where a teacher had successfully campaigned to install some Flow Hives for a program involving autistic children. The students had some experience with horses and dogs and were accustomed to having to keep the animals calm, but in this case the teacher aimed to help the students to learn to calm themselves through working with bees.

Before opening the hive to inspect your brood box, get your smoker ready and put on your protective gear. Although we've discussed this already on page 97, it's worth repeating that you should wear a good bee suit and gloves (also avoid wearing strong fragrances). Some bees are so gentle you hardly even need a suit, but it's better to be cautious until you really know what you're doing. Once you've suited up, collect your pre-prepared smoker, hive tool and any other equipment you may need (for example, a lighter, extra smoker fuel and a logbook).

Light your smoker and make sure the smoke isn't too hot (see the instructions on page 98). When you get to the hive, listen to the bees' hum. With experience you'll get to know what they usually sound like and become sensitive to shifts in the tone. Once you're ready to start your inspection, blow a few good puffs of smoke right into the hive entrance, then leave the smoker near the entrance so the bees keep getting a waft of smoke as you work, and it's within easy reach when you need it. It's normal for the bees to react to the puffs of smoke with a whooshing noise before they calm down.

At any stage during your inspection, if the bees sound agitated or start banging into you, stay calm, use a bit more smoke and then proceed gently. If they begin exhibiting more signs of aggression, you might want to wrap things up sooner rather than later. Put everything back the way it was and come back another time.

2 Opening up

Let's dive in! Gently remove the gabled roof and place it to one side. If your Flow Hive has roof locks securing the roof, you'll need to undo these first.

Use the sharp end of your hive tool to loosen the inner cover before removing it (the bees will typically have used propolis to stick it to the box beneath, so a little prying and levering at the corners is often needed).

You'll most likely see quite a few bees crawling on the underside of the cover. If there's no super and queen excluder on the hive, look for the queen (see below for instructions), because you need to make sure you don't separate her from the brood nest. If you can't see the queen, or can't confidently identify her, gently lean the cover against the hive close to the entrance so that if her royal highness is on it she'll be able to walk back inside.

Removing the super

If this is your first brood inspection, you're unlikely to have added a honey super yet. Otherwise, if it's an established hive with a super box on top, you'll need to remove it to get to the brood box. This can be done very easily using our Super Lifter invention.

Clockwise from top left The Flow Super Lifter; clamped to the side of the Flow Hive, a support leg drops down to take the weight of the super; the beehive swings open with minimal effort, allowing access to the brood box below.

The Flow Super Lifter

We have created an easier way to open a beehive. The Flow Super Lifter does away with the heavy lifting that might otherwise be a barrier to working with bees. There's still a little effort involved, but it's not back-breaking. Usually, a beekeeper has to manually lift a super that can weigh up to 30 kilograms (66 pounds) from the hive in order to inspect the brood. Add to this the awkwardness of lifting the box away from your body while trying not to squash bees. A further barrier is that the bees glue the boxes together with propolis, so levering the glued boxes apart can require considerable strength and skill.

We began this particular inventing journey shortly after the Flow Hive was finally on the market in 2015. There have been many attempts to make it easier to open a hive, some involving small cranes, rolling tables or drawers. There are existing designs for a drawer-based hive online and I made many variations of these. However, the bees were almost always angry when I slid the brood box out from under the super. At first I thought that the sliding action was annoying them, but finally I realised that as soon as the brood box moves backwards, the guard bees at the entrance become alarmed and quickly spread their alarm pheromone through the hive. We scrapped the drawer hive idea and went on to experiment with various other ways to easily access the brood.

The Super Lifter enables a beekeeper to crack the boxes apart and swing open their beehive with a minimum of effort. Firstly it will break the propolis glue binding the boxes together and then it can be used to lift one or two full-depth supers up and out of the way, allowing easy access to the brood box below.

If you don't have a Super Lifter, you'll need to do it the old-fashioned way. Take off the rear observation cover – this leaves you with a great point to lift from. Use your hive tool to loosen the corners above or below the queen excluder, then lift the box off. If it's too heavy you may need an assistant or you can remove some of the frames to lighten the load first. When you place the super down, stand it on one end, being careful not to tilt it to the side as nectar might spill out of any uncapped cells. The queen excluder will be stuck to the top of the brood box or to the bottom of the super. Take the excluder off, check for the queen on the underside, and leave it nearby or leaning against the hive. We recommend twisting the excluder rather than pulling it off so as to minimise abrupt movements that upset the bees. If you see a change in their behaviour it might be a good time to smoke them again.

Top left Smoke helps calm the bees.
Above Using a J-style tool. **Right**
Handle the frames gently. **Following
pages** In this frame we can see fresh
comb, bee bread in the bottom left,
and capped worker brood on the right.

If your bees are upset

Is there an intense and aggressive hum coming from the hive? Have you noticed agitated bees near the entrance? Are there bees bumping into you or trying to sting? Give them some smoke and wait a few minutes to see if they calm down. If not, there are a few reasons why your bees might be upset. Are you wearing a strong or unusual fragrance? Have you bumped the hive, or squashed some bees? Has the hive been open for a long time? Is it windy, cold or rainy?

If there are no changes to normal conditions, the issue may be within the hive. It could be queenless, or maybe they're hungry, or nectar from a new source could be impacting their mood. Like us, bee colonies have bad days, too, and some seem to become grumpier over time. Don't worry, each of these issues can be dealt with in a controlled way. Head to page 222 for troubleshooting advice.

3 Removing and handling frames

Now you're looking directly into the brood box. If the colony is well established, you'll see lots of bees at the top of the frames. The bees usually organise their brood nest with honey towards the outer edges of the box, followed by bee bread (fermented pollen) stores, then drone brood cells, with worker brood cells nestled in the central frames. To see what's going on in the brood nest, you need to remove frames and check them out one by one.

If the bees haven't built their combs straight within each wooden frame, it's going to be difficult or impossible to lift out individual frames without breaking the comb. This is called 'cross comb'. Don't panic, it can be fixed! For more info on how to deal with it, see page 236.

How to use a J-style hive tool to remove frames

The first frame is the most difficult to remove as there's not a lot of room to manoeuvre. First, look down between the frames and choose one that has little or no burr comb joining it to neighbouring frames. Give a few puffs of smoke to clear bees from the faces and ends of the frame. Then use the chisel end of your hive tool to apply some force between the frames to break the propolis that's connecting their end bars.

Now you can use the J-end of the tool to lever one end of the frame up high enough so that you can grasp it with your fingers. Keep hold of that end and repeat the levering process at the other end. Once it's coming up freely, you can slowly and gently lift it directly upwards out of the box with your hands, being careful not to roll any bees.

Make sure to disinfect your hive tool between inspections to reduce the chance of spreading pathogens from one hive to another. We recommend giving it a blast with a blow torch.

Handling foundationless frames

Assuming you're using foundationless frames in your brood box, the comb will often be attached to the frame only at the top. Be very careful when handling it in this case, as the comb can easily break. Avoid tilting the frame at an angle that would cause gravity to pull the comb away from the frame; instead move it from side to side and up and down. When you want to rotate the frame to look at the other side, you can swivel it 90 degrees horizontally, then 180 degrees vertically, then another 90 degrees horizontally. Sounds confusing? It's actually quite easy once you've seen it done! Check out our video link in the Resources section on page 276.

Tiny eggs in cells.

Uncapped cells with crescent-shaped larvae.

This regular pattern of capped worker brood indicates a healthy queen.

A patchy brood pattern could mean that something is wrong, or the hive isn't bringing in much forage.

Multicoloured bee bread stored in cells.

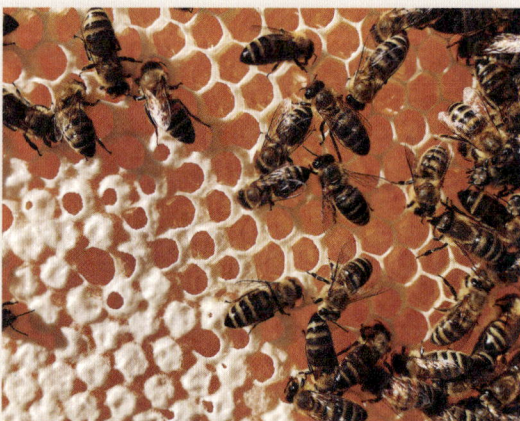
Capped honey and nectar stores.

Following pages Capped worker brood on the left, capped honey in the top right corner, bee bread on the right and some uncapped larvae in the bottom right corner.

Take a moment to admire the architecture of the comb. It's pretty extraordinary to think that this magnificent structure is the result of tiny creatures working together in the dark. Once you're finished inspecting the first frame, lean it against the hive or use frame rests to hang it on the side of the brood box. Remember, you want to make it easy for the queen to get back into the hive if she was on the frame you just removed. Make a mental note of where the frame was positioned in the box and which way it was facing so that you'll be able to put it back in the same place.

The second and subsequent frames will be easier to remove than the first, as you'll have space to move them sideways before lifting. They can be returned directly to the brood box as you finish inspecting each of them.

4 What am I looking for?

If this is your first inspection, your main aim is to get comfortable using the gear, opening the hive, lifting out frames and closing the hive again. Once you're feeling comfortable with all that, you're ready to start checking for indicators of your bees' health. It's a skill that builds over time as you get more familiar with bees and their intricate workings. Here are the main things you should be looking out for:

Developing brood – from eggs to capped cells

The primary thing you're hoping to see is worker brood, which indicates the presence of a laying queen. You will hopefully also be able to see eggs – they look like very tiny grains of rice. You'll need to use a flashlight, or if it is a sunny day you can turn your body so that the sunlight is coming over your shoulder and move the frame until sunlight hits the bottom of the cells (remember not to tilt it too much!). When all is well, you should be able to see an egg in the bottom of some of the cells. If you see multiple eggs in cells, or eggs on the side walls of cells, head to page 225 for troubleshooting.

As you look through the frames, you're hoping to see lots of capped worker brood in a fairly regular pattern, particularly in the central frames. You may also see drone brood, which has a domed wax covering that sticks out beyond the comb surface, in comparison to the worker brood cell capping, which is flatter though still slightly convex. The colour of the cells also varies depending on how new the comb is. Comb that has been used many times is darker than fresh comb. If you're lucky, you might even see a furry young bee emerging from a cell into the world for the first time!

Food stores

Something else you'll notice in a healthy functioning colony are stores of nectar and bee bread. Capped honey cells are also a good sign, as it means the hive has enough supplies to put some aside for the future. To a new beekeeper, capped honey can be confused with capped worker brood, but the airtight honey cappings are flatter and more translucent.

Pests and pathogens

Finally, look for signs of unwelcome guests or illness. One of the first signs of something amiss is a really patchy brood pattern, sunken dark cappings or discoloured larvae. If you see anything that looks off, don't panic! Check the troubleshooting guide on page 221 to determine what action needs to be taken.

5 Where's the queen?

You don't need to see the queen to know that she's there. If there are freshly laid eggs or developing larvae and plenty of capped worker cells, you can be pretty sure she's alive and well. Even so, it's always fun to practise queen spotting; as it is sometimes necessary to catch her (for example,if you're requeening), it's good to learn how to find her.

The queen will most often be found on a frame with lots of bees. Although she can be hard to spot quickly, her unique appearance and the way she moves are noticeable. Don't be discouraged if it isn't easy at first – she can be quite elusive.

Her majesty has a longer abdomen than the other bees and wings that are half the length of her body. Her legs are usually splayed at her sides and are longer, thicker and lighter in colour than workers' legs. She walks with a bit of a waddle, but she also moves much faster. Her thorax – the section behind her head – will very often be bald and shiny black, and the colour of her abdomen can vary from black to gold or anything in between. She'll often have fewer stripes than her workers.

The queen is sometimes surrounded by her attendants, which spread out around her like the petals of a flower. She'll leave them following in her wake when she's on the move. Rather than zeroing in on specific bees, taking a wider view of the frame and relaxing your eyes a bit can make spotting the queen easier, as you can notice changes in the pattern of movement more easily this way.

Some beekeepers like to mark their queen to make her easy to identify. If you want to mark yours, get to know the colour system that can be used to represent the year she was born so that you'll be able to tell how old she is at a glance (see more in the Resources section on page 276).

6 Closing the hive

When every frame has been inspected, try to put all the frames back in their original position. If you forget where the frames were or need to change the order for some reason, make sure there's space between the comb of each frame and the next. If the comb isn't very straight and there's anywhere without at least 4 millimetres (³⁄₁₆ inch) of space between opposing comb surfaces, the bees won't be able to service that area and small hive beetles could lay lots of eggs in it. You definitely don't want your hive to become a beetle nest!

Pack the frames together tightly towards the centre of the box, leaving any excess space at the edges. Then carefully replace the inner cover and the roof (plus your queen excluder and super if you removed them at the start).

Clockwise from top Inspecting a brood frame; the Flow Beekeeping Caddy is a handy piece of kit at inspection time; a smoker will encourage bees to move away; a queen bee and a worker.

Game of Combs: there can only be one

When a hive is preparing to swarm, the workers will begin raising multiple new queens in peanut-like cells. Mature queens kept enclosed by the workers in their cells will 'quack' to signal to the colony that they're ready to replace the queen if she leaves as part of a primary swarm or there is a secondary afterswarm.

The first of them to emerge will make a high-pitched 'tooting' sound to let the colony know that she's out. This induces the workers not to uncap any other queen cells, which reduces the chances of a deadly battle royale. If two queens emerge at the same time, the fight between them can last up to 15 minutes, and the outcome is fatal for one of the participants.

If the newly emerged monarch doesn't leave with an afterswarm, she will typically locate and kill other queens by stinging them while they're still trapped in their cells. And when a mated queen is introduced to a hive by a beekeeper, she sometimes makes a loud high-pitched piping sound that announces her presence. Some researchers think this may also serve as a battle cry, challenging any rival queens that might be already present.

The queen's longer abdomen will often be hidden by other bees.

When trying to spot the queen, relax your eyes and look for a different pattern of movement.

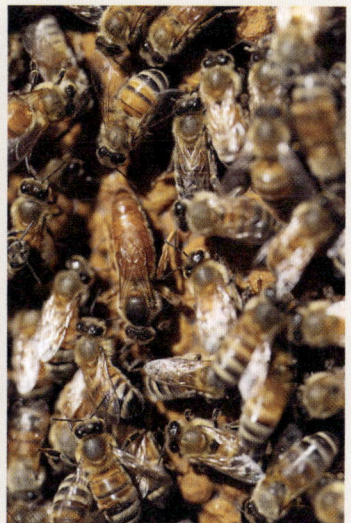

Bees will often turn to face the queen as she moves across the frames.

When things are not as they should be

Here are some of the main warning signs to look out for in your hive. For detailed information, go to page 221.

Your hive is queenless

As we mentioned previously it can be hard to spot the queen. Sometimes, that's because she's not there. If a queen has been absent for a while, the cells will have no new supply of fertilised eggs. Your first clue that the hive is queenless is if there's no sign of eggs or young larvae, or if there are multiple eggs in cells (this could mean the workers are laying, or sometimes a newly mated queen can lay multiple eggs as she gets the hang of things).

The population of new bees will dwindle quickly without a queen, which affects the number of nurse bees, as they have nothing to do without larvae to feed. Instead, they'll turn to other work, such as gathering nectar and pollen and making honey, so the stores of these essentials will increase. No queen means no queen pheromone. When a laying queen goes missing or dies, the lack of queen pheromone will trigger the colony's instinct to raise a new queen, but this doesn't always succeed.

If your hive appears to be queenless, it's usually best to add a frame with brood comb from another hive. If there are fertilised eggs in the cells, the bees will usually start raising their new queen themselves. Within a couple of days they'll tear down the cell wall around an existing larva that has just hatched, build a queen cell around it and feed the larva royal jelly until it's ready to pupate (see page 65).

Queen/supersedure cells or swarm cells

Queen cells are elongated peanut-shaped cells used for raising new queens. There are three reasons why a colony might make queen cells: the hive has suddenly become queenless; the queen falls ill or fails and is replaced naturally (supersedure); or if the bees are preparing to swarm.

Queen cells located in the middle of a frame are usually emergency or supersedure cells. If there are lots of bees and you find queen cells at the edges or bottom of frames, they are most likely preparing to swarm. This typically happens when floral resources are abundant and the colony starts getting too big for the hive. When they swarm, the original queen will leave the hive accompanied by half the colony before a new queen emerges. It's best to take action to prevent swarming before it happens. (See page 202 for info on how to prevent swarming.)

Pests and bee health

The presence of a few pests is normal, but too many unwelcome visitors interfere with the growth and health of the hive. Depending on where you live, different pests might be more prevalent. Symptoms of illness vary. Check the larvae and brood cappings as they are often the first places disease is visible. Changes in colour or smell or the presence of fungus are warning signs of a colony's ill health. Any unusual changes in the hive should be investigated. Some diseases need to be reported and may have specialised treatment requirements.

🐝

Checking for varroa

Varroa mites are a very destructive pest, and it's necessary to monitor regularly for them. In many regions, beekeepers use the screened bottom board to estimate mite levels. Washes are a far more accurate method. For more information about dealing with varroa, see page 244.

No stores in the brood box

Bees store honey and bee bread in the brood nest to provide food for the colony. Without a good diet, bees are less able to fight off pests and disease. The queen will lay with confidence only if there's sufficient food for the new generation. When honey stores are low because of a lack of nectar in your area, consider feeding the hive. (See page 207 for instructions on how and when to feed your colony.)

An emergency or supersedure cell in the middle of a frame.

Following pages Note the drone brood with convex cappings on the left-hand side of this frame. You can also see some capped honey at the very top right.

A swarm cell at the edge of a frame.

When and how to add a honey super

Patience is an important part of beekeeping. Each colony works according to its own timetable and will be impacted by many factors, including the strength of the hive and the availability of nectar. Sometimes your hive will be ready for a super just a couple of weeks after you install bees; in other cases it can take an entire season before it's time. On average it will take at least a few months, all depending on your bees, your climate and the yearly cycle.

Timing is everything

Once all the frames in the brood box have been built out with comb, the box is full of bees and there is plenty of brood, indicating that the colony is expanding, it might be time to add one or more supers to your hive. A lot of uncapped cells filled with glistening nectar indicate that an abundance of forage is being brought in, so the bees may be ready to start using extra space for honey storage. Luckily for us, they often store more than they need, and we can harvest the surplus.

Consider the season you're heading into: there should be good availability of nectar locally for the foreseeable future. Adding a super too soon can be detrimental to the hive, because a small colony may be unable to regulate the temperature in the additional space or to guard the super from pests. (For more information on what to expect in each season, go to page 195.)

Preparing and adding your Flow Super

Once you're confident that your hive is ready, you need to ensure the Flow Frame cells in your Flow Super are correctly aligned. Because the components may have shifted during transit, it's important to reset them by inserting your Flow Key into the uppermost slot and turning it: this pushes the moving parts downwards to form hexagonal guides for the bees. Install the Flow Frames in your super, packing them together towards the centre with the harvesting end facing the back where the removable cover is. You can adjust the screw at the top end of the Flow Frames to keep them from moving back and forth. Line up the viewing end of the frames so that there are no gaps where bees can escape. You're now ready to put a queen excluder on top of the brood box and place your super on top of that.

It's usually preferable to have the harvesting end of the super at the opposite end of the hive to the bees' entrance below the brood box, as you want to be able to stand at the back of the hive to harvest, not at the front blocking the flight path. As you slowly lower the box onto the hive, gently smoke the bees to encourage them to move out of the way so they don't get squashed. Line the sides of the super up with the brood box. When you're happy with the positioning, put the inner cover on top of the super, close the feeder hole and secure the gabled roof.

Encouraging the bees to move on up

Some beekeepers end up harvesting honey within just three weeks of adding a super; most will have to wait quite a bit longer. For the bees to store honey in your Flow Frames two conditions need to be in place: a thriving colony (you should see lots of bees when you open the viewing windows) and a good nectar flow. These two things don't always align, so it could be months before you see much activity on your Flow Frames.

We never wax our Flow Frames, but if you're getting impatient you can add some beeswax to the Flow Frames. The easiest way to do this is to scrape some burr comb off the top of your brood frames (no need for any honey) and press it into the Flow Frame surface. Put the frame in the super next to an observation window so you can then enjoy watching the bees recycle the wax for completing the cells. Once they have sealed the joins in the Flow Frames, they will wax the cell walls before starting to store nectar. Some beekeepers like to melt beeswax onto the Flow Frames. While this can be done, be careful not to pour wax into the Flow Frame cells or you may find them very difficult to operate later.

A good nectar flow and a lot of bees in the box are still necessary before you'll begin to see honey stores developing. If you have multiple brood boxes or supers, we recommend changing the configuration to a single brood box and a single super on top. For faster results, we recommend allowing the bees to make a good start on the Flow Frames before adding additional supers or brood boxes; however, seek local knowledge, as in some areas of the world it's common practice to run two brood boxes before adding any honey supers. For example, master beekeeper Fred Dunn in Pennsylvania finds that he gets the best results from using a deep brood box at the bottom with an ideal (shallow) brood box on top and then the Flow Super on top of that. Another trick is to switch the positions of the brood box and super, thus forcing the bees to travel from the entrance through the super to get to the brood. After a week or so the boxes can be put back into the usual order. Also note that sometimes genetics are the problem: you could have two hives side by side with one bringing in lots of nectar while not much is happening in the other. This is one example of the benefits of having at least two hives.

Keeping records

Logging your beekeeping activities is a good thing to do. It's especially useful if you're running multiple hives, and it can also help you learn more about your bees and get better at beekeeping.

Record keeping can be done using a physical logbook or with an app. Each time you check your hive you can make a note of observations such as the quantity of bees, the quality of the brood pattern, whether you saw eggs or the queen, how much honey is being stored and the presence of any pests. You can also record any actions taken. Over time you'll build up a record of information that can be used to guide future planning or decision-making.

Top right The bees have begun capping the cells at the top of this Flow Frame. **Middle left and right** Note where the bees have waxed the gaps between the Flow Frame cells. **Below** The Flow office apiary.

Top things you're looking for in a brood inspection

As well as admiring your bees and their interesting behaviours, ask yourself:

- **Are they queenright?** Look for eggs, larvae and capped brood. It's fun to spot the queen if you can! See the troubleshooting chapter on page 221 if you think your queen might be absent or ill.

- **Are they healthy?** Check for signs of pests and diseases. Depending on the season and how long it's been since your last inspection, it may be time to monitor for varroa mites. If anything seems amiss, refer to the troubleshooting chapter for more information.

- **Is the population increasing or decreasing?** Understanding the momentum of a colony is a valuable skill to develop. Signs of a growing colony include numerous nurse bees or young bees, the presence of abundant brood at all stages, and fresh comb being drawn. Patchy capped brood and fewer young bees could indicate that numbers are waning. You may need to take action accordingly, such as adding or removing a super.

- **Are they at risk of swarming?** Look for queen cells, especially at the bottom of frames, and take action quickly if you find any. The hive could start preparing to swarm if it's very crowded or if there's an abundance of nectar in the brood area (for information on honey-bound hives see page 204).

- **Do they need feeding?** Do your bees look like they have enough stores of nectar and pollen? Is a dearth expected in the near future? For more info on nutrition and feeding, go to page 207.

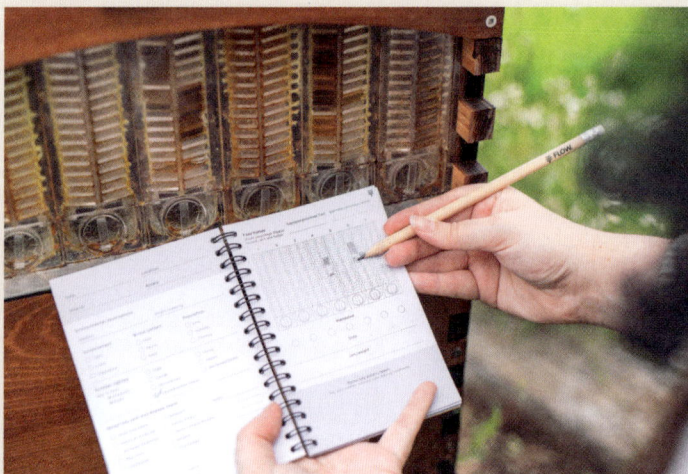

Some beekeepers use a logbook for recording observations and events in their apiary. **Opposite page** A page from our logbook gives you an overview of things you might look for during inspections.

Date: _____   Location: _____

Hive ID: _____   Apiary: _____

Environmental observations

Weather: _____   What's flowering: _____

Temperament	Brood pattern	Population
⬡ Calm	⬡ Poor	⬡ Low
⬡ Lively	⬡ Patchy	⬡ Average
⬡ Aggressive	⬡ Solid	⬡ Thriving

Number sighted

Key: **L** (low)
    M (medium)
    H (high)

⬡ Eggs   ⬡ Honey

⬡ Larvae   ⬡ Nectar

⬡ Drone brood   ⬡ Bee bread/pollen

⬡ Capped worker brood

Brood box pest and disease check

⬡ Small hive beetle	⬡ Sacbrood	Notes: _____
⬡ American foulbrood	⬡ Varroa mite(s)	_____
⬡ European foulbrood	⬡ Colony collapse disorder	_____
⬡ Wax moth	⬡ Nosema	_____
⬡ Chalkbrood	⬡ Tropilaelaps	_____

Queen

⬡ Sighted	⬡ Colour: _____	⬡ Swarm cells
⬡ Laying queen	⬡ Virgin	⬡ Emergency cells
⬡ Marked	⬡ Queen cups	⬡ Supersedure cells

Bees and space

Lachlan Thompson, Victoria, Australia

Professor Lachlan Thompson once sent bees into space, having designed an experiment involving spiders and pollinators on the Columbia space shuttle. He now keeps Flow Hives at his own space observatory, where he applies his scientific knowledge to beekeeping in the hope that monitoring beehives and their relationship to the climate will be a key factor in helping to restore our planet. To see more of Lachlan's story, check out the Resources section on page 276.

'The ancient Greeks thought that honey was divine and a gift from the gods. And it is. Well, it's a gift from the bees.'

MEET THE BEEKEEPER

Beekeeping with mum

Janabel, Gaudy and Lumiere, New South Wales, Australia

Janabel and Gaudy are beekeeping sisters who had a family bonding experience when they introduced their mum, Lumiere, to the fascinating world of a honey bee colony. To see more of Janabel, Gaudy and Lumiere's story, check out the Resources section on page 276.

'To be able to come home and see my family and do inspections together, it's just a really nice thing to do on the weekends.'

Scan here for videos relating to this chapter

6

The honey harvest

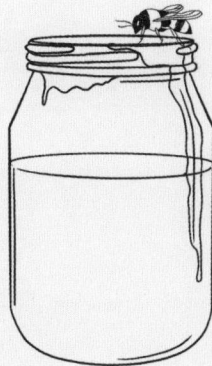

No matter how many times we do it, every harvest from a Flow Hive still feels a little bit like magic! Watching liquid gold pour straight into a jar, while the bees are left to just bee, is an experience that's especially fun to share with friends and family.

When you eat honey from your own hive, you're tasting the unique essence of your botanical surroundings that is distinct to the region's latitude, soil and climate. What's more, every bite evokes something that only a beekeeper can feel – the memory of the care you've given your bees and an appreciation for the work they've done to create this amazing substance.

To us, as well as being something delicious that we love eating and giving to people as a gift, honey is something to revere, to behold, to collect. It's a sweet representation of the extraordinarily interconnected nature of life all around us.

A (very) short history of honey

Historians and anthropologists reckon that humans have been eating honey since the Stone Age. It's been theorised that energy-dense honey – more than a just a sweet treat – may have played a part in the development of the human brain.

Honey has been an ingredient in food and drinks since ancient times. By around 9000 BC, 'honey water' was being fermented to create mead, which is believed to have been one of the first ever alcoholic drinks.

The oldest stash of honey ever found was discovered in a ceramic jar buried in an ancient noblewoman's grave in the Republic of Georgia. It's thought to be around 5500 years old. Until this discovery, honey from ancient Egyptian tombs was considered the oldest in the world. The ancient Egyptians were keen beekeepers, transporting log hives up and down the Nile, and they also used honey as an ointment for wounds, as an offering to the gods and for embalming the dead. When King Tutankhamun's tomb was excavated in 1922 the archaeologists found 3000-year-old jars of honey buried with the king. Tasting it (which was brave of them if you ask us), they found it was still edible.

Honey is embedded in many religious and cultural ceremonies. Buddhists often donate honey to monks to emulate the story of a monkey who gave honey to Buddha himself. The Quran mentions honey's healing properties. In Hinduism, honey is a food of the gods and one of the five elixirs of life. Apples are dipped in honey during the Jewish festival of Rosh Hashanah to symbolise a sweet new year. Honey's value has been monetary as well as nutritional and symbolic: ancient Romans used honey to pay their taxes, and even today it forms part of the 'bride price' in some African communities.

Honey flavours

In our subtropical region on the east coast of Australia, the bees create a great variety of flavours throughout the year. In the spring time we get honey that ranges from a light straw colour to almost clear, bursting with flavours that are so vivid it's like eating the flowers themselves. In summer we get the iconic Australian eucalypts with their well known taste, and in winter when the rain sets off the paperbark tree blooms we get fluoro yellow nectar coming into the hive that produces a brown-red honey reminiscent of slightly burnt toffee. And then there are all the other flavours in between. You can't always pinpoint where they come from – some taste like lollies (candy), others are bitter and make you screw up your face, and there's even one so strong that we say it tastes like possum piss! When we have honey tastings, it's always fun to throw in a few curveballs to really get the conversation going.

Fine flavours

When we invented the Flow Frame harvesting system, we were primarily trying to make things easier for the beekeepers and the bees. We didn't realise that the quality of honey from a Flow Hive would be superior in two ways.

Firstly, the Flow Frames allow you to easily harvest monofloral honey because the bees tend to store different nectars in different frames. You can often see the varying honey colours in each frame through the rear window.

Secondly, to our surprise, the honey from Flow Frames tastes very fresh, with particularly strong floral notes. Not trusting our own biased tastebuds, in 2016 we asked the University of Queensland to do independent taste tests and mass spectrometer analysis. The results backed up our opinion and that of many other beekeepers: Flow Hive honey does taste livelier and fresher than honey extracted conventionally with a centrifuge. We think this is because commercial honeys are often flash heated, a process that destroys many nutrients and subtleties of flavour, whereas honey from Flow Frames is unfiltered, unheated, unmixed and much less oxidised.

We didn't set out to be able to isolate honey flavours, but one of the extraordinary things about the Flow Hive is the way you can harvest frame by frame and therefore experience so many unique flavours, rather than mixing them all together. It's such a joy to taste what the bees bring in and it's also a reflection of your local landscape. There are about as many different flavours and colours of honey as there are nectar-producing flower species in the world, and it's a never-ending game of join-the-dots to work out where the beautiful flavours are coming from.

To help identify which floral source a honey comes from, aside from noting the aroma, which is often a giveaway, you can taste the nectar from likely flowers. Whack the flowers on your hand to extract the nectar droplets. If you'd like to learn about identifying honey flavours, check out the Resources section on page 276.

When to harvest

When I used to harvest honey in the conventional way, I would wait until 20 hives in an apiary were all more or less ready to go at once, as batch processing makes the whole ordeal easier.
If some hives were doing particularly well, I would add more supers to them. For the most part though, a number of hives would sit there full while I'd wait for the others to catch up. Using Flow Frames allows you to harvest a single frame as soon as it's ready, making space for the bees to store more honey. So instead of trying to store honey in more supers on hives while you wait for the time to be right for batch processing, you can harvest more regularly and store the honey in jars on the shelf. This way you don't need to have and maintain lots of spare supers. As a bonus, if you're unsure whether the bees have enough honey for a shortage of forage ahead, you can easily harvest just a single frame or even part of a frame and leave the rest for the bees.

— **Cedar**

It's a pretty exciting moment the first time you see your bees storing honey. You can watch them starting to fill the cells through the side observation window on the Flow Hive, and if you take a close look through the end window you might even see them depositing nectar with their tongues! The time to harvest is when there is enough capped honey stored in the super and you are confident the bees will not be left short. In beekeeping, patience is a virtue that your bees will thank you for!

The windows of the Flow Hive let you see if the bees are bringing in fresh nectar and building up their stores, or if they are a bit hungry and eating into their larder. Watching over time will give you a good idea if they're filling and are likely to have capped the Flow Frames all the way through, or if they're hungry and there could be sections of the comb that have been uncapped and consumed. If you're unsure whether they have enough for you to take a full harvest, a feature of the Flow Hive is that you can easily harvest just a small amount and leave the rest for the bees.

In a temperate climate with distinct seasons, the honey harvesting window will likely start in mid- to late spring and extend throughout summer and possibly into autumn; however, seasonality can vary quite a bit from year to year, and you'll learn to understand what's happening in your hive and be able to spot the signs of when it's ready to be harvested. If your hive is only just getting started, you'll have to wait until your colony has established brood comb, built up their numbers and put away a decent store of honey before you can take your first harvest. It may be a full year before this happens. (See page 152 for info on how to encourage your bees to start filling the Flow Frames if they're slow to do so after you've added a super.) If there's not much nectar around and no flowers are expected soon, or there's a long spell of bad weather approaching, or you're about to head into winter, you shouldn't harvest.

To determine whether your honey is ready to be extracted, check the observation windows in the Flow Super. The rear observation window gives you a cross-section view of the bees filling the cells. When the honey is ready and the bees are happy with the moisture content, they will put a wax capping over the top of each cell. This is like putting a lid on a preserving jar to store the produce for a time of need. Luckily for us they are so productive and often store more than they need!

The side observation windows are useful, too, to see how capped the outer frames are and get a general gauge on whether the honey stores are increasing or decreasing.

How bees make honey

Flying from flower to flower, a female forager bee uses her long proboscis (a straw-like tongue) to collect nectar. This sweet substance is typically 70–80 per cent water, with the remainder composed of sugars and other compounds. As she flies, enzymes in her 'honey stomach' (a specialised compartment separate from her digestive system) begin breaking the complex sugar molecules into simpler ones and dewatering the nectar.

When she returns to the hive, she passes the partially processed nectar mouth-to-mouth to one of her younger sisters, who then chews it for about 30 minutes, adding more enzymes. She'll then pass it to another bee, and this is repeated several times. More water evaporates during this process, and each bee involved contributes to converting the nectar into a thicker, more stable form.

The concentrated nectar is deposited in a hexagonal honeycomb cell. After that, more moisture is evaporated off by the fanning of bees' wings until the honey is 'ripe', meaning that the water content has been reduced to 17–18 per cent. Now that such a high sugar concentration has been reached, osmotic pressure prevents microbial growth, meaning that the honey won't ferment or spoil. The bees cover the cell with an airtight wax capping.

It's difficult to measure this exactly, but it's commonly believed that a single worker bee collects enough nectar to make about a twelfth of a teaspoon of honey over her entire lifetime.

The worker bees generally start filling the centre frames first and work their way out to the extremities. In times when nectar is scarce, the nurse bees will usually take honey from directly above the brood nest. So if you notice the bees are uncapping cells and leaving a patchy honey pattern, you may also see the centre frames have a half-moon of honey missing, even though the rear frame view is showing capped honey. For this reason, when the bees are depleting honey stores it's best to harvest only the frames toward the edge of the hive or not harvest at all, leaving the stores for the bees.

On rare occasions, the bees won't fill the frames all the way to the end. This can be because of their genetics, or perhaps because too much light is getting in the rear door. In this case, it might look like there's no honey, but really, the super is full and harvestable. If you think you might have either of these issues, fully inspecting your Flow Frames is a great way to confirm.

If honey is harvested before enough of it has been properly capped, it may have a high moisture content which can cause it to ferment. Known as 'baker's honey', fermented honey has a sharp taste. It can be used in cooking and mead production, but you're not going to want to put it on your toast. As a rule of thumb, a moisture content of around 18 per cent will mean that the honey will keep almost forever, as long as the jar has an airtight lid. If the moisture content is higher than this, fermentation will likely occur at some stage, depending on how much yeast is present in the honey. A honey refractometer can be used to measure the moisture content. However, if you make sure that your Flow Frames are at least 80 per cent capped before harvesting, your honey should be fine.

Above left These Flow Frames look like they have plenty of honey in them. **Above right** Cells are fully capped and ready to harvest.

How much honey will I get?

The amount of honey your hive produces will depend on many factors, including colony strength, local climate and available forage. This will change from year to year. If your Flow Super is full, you may expect to harvest approximately 3 kilograms (6.6 pounds) per Flow Frame (or even more if the bees build each frame out to maximise storage).

Harvesting from a full super multiple times in one season is possible if there's a good nectar flow. However, even if you have a full super, it doesn't mean you should harvest all your honey at once. As mentioned above, you've always got to consider how much the bees will need for themselves.

One full Flow Frame will produce about 2 litres (8 cups) of honey, give or take. If you're using multiple small jars rather than a couple of very big ones, having some extra jars on hand is usually a good idea. One of the benefits of a Flow Hive is it lets you take as much or as little honey as you want at a time. It's easy to just harvest a frame or two and leave the rest for the bees, or even harvest just a part of a frame by inserting the key only part way in before turning.

Pre-harvest checks

Below are three things you should do to ensure your harvest is successful.

Levelling your hive

Levelling your hive is an important step and should be done each time you harvest. Even experienced Flow Hive beekeepers sometimes miss this step and run into issues. It's common for the soil to sink under the weight of the hive and change level. It's important to have a 3-degree slope towards the honey harvesting end, and that the hive is level in the side-to-side direction. The Flow Hive 2 and later models have integrated levels at the side and the rear: simply adjust each corner of the hive until the level bubbles are central. If you have an earlier model you will need to level the surface that the base sits on, as the slope is built into the baseboard.

Setting up the tray or slider

If you have our integrated pest management base, make sure the tray is in place. If you have the classic base, make sure you put the corflute slider in the upper position. Through the harvesting process, you might find honey spilling inside the hive (many factors can contribute to this; if it is causing a problem see page 248). Bees are great at cleaning up spilt honey in the hive, but if any drips down through the screened bottom board and onto the ground it's likely to attract robber bees. For this reason, it's important to have your tray or slider in place.

Above, from left A completely capped Flow Frame; the built-in spirit level in a Flow Hive base ensures the hive is at the correct angle for harvesting. **Opposite page** You can insert the Flow Key partway to harvest only a section of a Flow Frame.

A tip about jar coverings

It's a good idea to have something available to cover your jar. Often you can harvest without the bees even noticing your jar filling up behind the hive. Other times, the bees may be hungry and on the lookout for anything sweet. If bees are coming for the honey, cover it up so they can't get into your jar. A piece of kitchen wrap or beeswax wrap will suffice. Put it over the top of the tube and jar, being careful to seal up any cracks that could allow a bee in.

Some people have made elbows for the Flow Tubes so that they can run tubes into holes in the lid of their jar or bucket. If you build a system like this, please ensure that the jar or bucket can breathe, because if air can't escape from the container while it is filling then the honey will back up and spill into the hive.

You can cover the Flow Tube with a beeswax wrap to prevent bees from robbing during the harvest.

Remove the end caps cover.

Remove the end cap from the top of the frame.

Insert the Flow Tube into the base of the frame.

Insert the Flow Key into the lower slot at the top of the frame.

Turn the Flow Key. Honey will flow out into your jar.

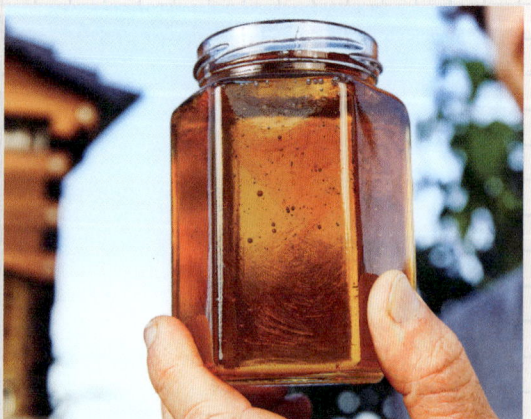

Liquid gold!

Flow Hive harvesting step-by-step

You will need:

- beekeeping suit or veil
- jars
- harvesting shelf brackets if you've got them, or something else to rest your jars on
- Flow Tubes
- kitchen wrap or wax wrap to cover your jar in case the bees are in a robbing mood
- Flow Key

Now let's go through the steps to a successful harvest one by one.

- Put on your protective gear.
- Choose your jar size.
- Attach your shelf brackets to the hive at the required height to suit your jar size and use the rear door as the shelf (it's best if the jar is sitting just below the outlet). If you don't have the shelf brackets you may need to find something to prop your jar up.
- Take a moment to appreciate all the work the bees have done!
- If you've got a Flow Hive Classic or Hybrid with a corflute baseboard, put the corflute in the top slot.
- Remove the end caps from the frame you're going to harvest from.
- Insert your Flow Tube (tongue end first, to prevent honey spill).
- Place a jar under the tube. Cover with kitchen wrap or wax wrap.
- Insert the Flow Key into the lower slot at the top of the Flow Frame.
- Turn the key! If it's difficult, try inserting and turning it in increments.
- Watch as the honey starts to flow.
- When the honey has almost stopped flowing into your jar you should reset the Flow Frame. Turn the key 90 degrees and slide it out. Now insert the key into the upper slot at the top of the frame and turn it into the vertical position. Make sure to leave it in this position for a minute or two to make sure all the Flow Frame cells return to the hexagonal position.
- You can now remove the tube and replace the caps.
- Congratulations on a successful harvest!

For more detailed instructions and a video demonstration, head to the Resources section on page 276. If you have any issues with your Flow Frames, check out the troubleshooting chapter on page 221 or contact our support team via the website.

After you've harvested, you will notice the bees are already at work chewing away at the wax cappings, recycling the wax to rebuild the cells so they can start refilling them with nectar. If you keep a logbook, you can note the harvest date and then see how long it takes for the bees to replenish the honey stores. This type of record keeping can be really helpful over time as it illustrates nectar flow patterns throughout the year.

Storing your honey

If honey is stored in an airtight container away from heat and light, it will remain edible far beyond your lifetime. Its amazing stability is thought to be due to its high sugar content and the presence of a particular enzyme that prevents bacterial growth. If your honey is left open to the elements it can absorb moisture from the air, which may eventually dilute it and lead to fermentation.

If your honey is runnier than you'd like, keeping it in the fridge can make it much thicker. Cooling honey can encourage crystallisation (also known as candying), which some people prefer. If honey does become crystallised, it's still perfectly fine to eat. You can quickly return it to its normal state by placing the sealed jar of honey in hot (but not boiling) water to help the crystals dissolve.

Honey and health

For thousands of years, people have known that honey can be used to treat wounds and skin problems. Hippocrates himself (460 to 375 BC), the most renowned physician in antiquity, prescribed honey for a number of conditions, including the treatment of wounds. Honey's unique healing qualities come from the complex interplay of more than 200 compounds, making it virtually impossible to replicate synthetically – there's just no substitute for the real thing. Honey's strong antibacterial properties, high acidity, antioxidants and hydrogen peroxide content make it an effective topical treatment for wound healing. It's also been used in cosmetics and skin balms for millennia.

> In ancient Greece, it was believed that feeding honey to a child would endow it with wisdom. Legends told of Zeus, king of the gods, being raised by nymphs called Melissae who fed him honey. It was even said that bees deposited honey directly into the infant god's mouth. To this day, the Greek word for honey is μέλι (méli). In the modern world, honey is not recommended for babies under 12 months old, as it may increase the risk of infant botulism.

Not only good for wounds, eating honey regularly impacts more than just your tastebuds: it has anti-diabetic qualities, and can also help set up your stomach for ongoing health. Honey is a potent prebiotic that can boost the levels of healthy bacteria in the gut, providing positive benefits for our immune systems. Honey can also help protect against gut disease, as the bacteria that feed on honey produce protective short-chain fatty acids. Many studies have shown that gut health is linked to overall wellbeing, with the health of your gut impacting every organ in your body.

Honey affects our brains too. Some researchers claim that the bioflavonoids and polyphenols present in honey have a significant neuroprotective effect, supporting brain function long-term and protecting it from degradation that can be caused by eating refined sugars. One day, we might be able to definitively say that eating honey instead of refined sugar makes you smarter as well as healthier! (Note: Although honey is much better for you than refined sugar, as with anything there's a limit to how much of it you should eat as part of a balanced diet.)

On certain mountainsides in Nepal and Turkey, there are species of rhododendron whose flowers contain compounds called grayanotoxins. Honey produced from the nectar of these flowers is called 'mad honey' because eating it can result in hallucinations. People brave dangerous heights and stings from the world's largest honey bees to collect this hallucinogenic honey, and there are cultural traditions surrounding its use for medicinal or ritual purposes. We certainly don't recommend that you try mad honey, because overdosing on grayanotoxins can lead to serious health complications including cardiac arrhythmia and respiratory failure.

A note on *Leptospermum*/manuka honeys

Leptospermum is a genus of flowering plants, more commonly known as tea trees or manuka myrtles. Australia is home to 84 of the 87 known species worldwide, and one species is found in New Zealand. Nectar from their flowers produces a thixotropic honey, which means it doesn't run or flow unless agitated. This is likely due to higher levels of proteins. Because it is solid at rest, thixotropic honey is difficult to remove from Flow Frames (or conventional frames for that matter). We have found when there is a 50/50 mix of thixotropic and non-thixotropic honey in a frame, the honey will flow, and you will see jelly-like globules coming out into your jar. Higher concentrations, however, may set in the frame and not flow. Opening and closing the cells repeatedly can help a little, as this has an agitating effect. See page 252 for advice on what to do if your Flow Frames are full of a thixotropic honey; for example, manuka and/or jellybush in Australia, ling or heather honey (from *Calluna vulgaris*) in Europe or grapefruit honey (from *Citrus paradisi*) in North America.

Other hive products

While honey is just about everybody's favourite thing that bees make, there is a range of other useful things you can get from your hive.

Beeswax

Humans have been using beeswax for many purposes for thousands of years - in candle-making, as an ingredient in cosmetics, as a polish or cleaner or lubricant, and in the arts. It's been found on pottery shards dating back to the 7th century BC, probably used for waterproofing.

Although wax is transparent when excreted from bees' abdomens, it becomes opaque as they chew it. Contact with pollen and propolis then turns it into a typically creamy yellow or gold colour, becoming deep bronze or almost black in very mature combs.

You can collect beeswax from your Flow Hive by cutting out some of the comb from the edges of your brood box. You could also provide another area for the bees to build their natural comb by removing the plug in the inner cover to let them into the roof cavity, or adding another super for the collection of honeycomb. You can then crush the comb to separate wax from honey.

Cleaning the wax is a two-step process. First, place the wax in a pot on the stove with some water and gently heat it until it melts. Impurities (also known as 'slumgum') should sink to the bottom of the pot. When you pour the wax into a tray, try to leave the slumgum behind. After the wax has cooled and solidified you will have a nice block of yellow beeswax for your craft projects.

Honeycomb

A thick piece of honeycomb straight from the hive is one of the world's greatest culinary pleasures. Serve it with a cheese board, over pancakes or porridge, with Greek yoghurt and fruit, as an edible decoration for desserts, or on a thick piece of warm bread with butter. Stu doesn't mind eating it straight from the hive tool.

Check there are no bees on the comb when you eat it! I often get carried away tasting all the different honeycomb flavours from around the hive while beekeeping. Once I ate a piece with a bee on it – ouch! Poor bee...

—Cedar

Harvesting honeycomb requires the same essential preparation as a hive inspection. The only difference is that you will most likely want to take the frame to a separate place (preferably indoors) to remove the comb, so the bees won't bother you while you're working. One advantage of foundationless frames in the brood box is that you can easily cut out a piece of honeycomb from a frame and put the frame straight back in the hive for the bees to repair. You can find a link to a video demonstration of honeycomb harvesting in the Resources section on page 276.

Propolis

Propolis is an interesting substance that's often overlooked by beginner beekeepers. The Greek word propolis is derived from the words 'pro' (in front of), and 'polis' (the city), referring to its role in defence of the hive. Bees use this antimicrobial 'glue', made from tree resin mixed with wax and pollen, to seal gaps or cavities, glue hive parts together, repair holes and provide a hygienic inner environment for the colony. Studies have shown that propolis helps to make the colony more resistant to pathogens.

In cold climates, autumn is peak propolis production time, as bees get busy sealing cracks in the hive in preparation for winter. However, some colonies have the propensity to collect far more propolis than others, and this is thought to be a tendency that is passed on through genetics.

In colder climates, despite the best efforts of guard bees, sometimes a mouse or other small predator might make its way into a hive. When this happens, the intruder usually doesn't last for long! About five bee stings later the mouse will be dead. But the bees now have a body on their hands. They can't leave it there because it will start to rot and contaminate the nest, and it is too heavy for the bees to move it themselves. So they encase it in propolis, creating a mummified mouse.

In our hives, propolis is usually sticky to the touch and dark brown in colour, but it can also be rock hard and coloured green, red, black or white, depending on temperature and the source of the resin. When you're inspecting your hive, you might sometimes see large amounts of propolis at either end of the top bars of the brood frames. These are actually storage piles that the bees can access whenever they need to. Propolis can also commonly be found holding down your inner cover, holding the super to the brood box, and sticking all the sidebars of your brood frames together. Propolis is the main reason beekeepers need to pry things apart with a hive tool!

It has been long thought that propolis has health-giving properties for people and it has been used for wound healing throughout the ages. It is also high in antioxidants, and some of its claimed benefits for humans include helping to fight bacteria, viruses and fungi, as well as having anti-inflammatory and immuno-modulating effects.

Propolis can be removed from the hive by simply scraping it off using a hive tool. Some beekeepers use a propolis tray or mesh-like trap to collect it. Remember, the bees need propolis too, so just like with honey harvesting, it's a good idea to leave them enough for their needs. Although some beekeepers chew propolis straight from the hive, others prefer to clean it first. This can be achieved by soaking propolis in water for three to seven days, which allows any debris or dirt to float to the surface. Once clean, the propolis can be air-dried, before being ground into a powder or dissolved in alcohol.

Royal jelly

Royal jelly has long been a highly prized bee product. The ancient Greek philosopher Aristotle realised the role of royal jelly in raising a queen, making the connection between her longer life and physical strength. In ancient China dynastic families believed it could extend their longevity and enhance their sexual potency; Cleopatra used royal jelly as part of her beauty routine; and India's maharajas considered it essential for boosting health.

Some studies are showing wisdom in these practices. Early promising results of royal jelly's potential benefits include relief from the symptoms of menopause, reduction in cholesterol, regulation of blood sugar levels, an increase in the speed of wound healing, and lowering blood pressure.

However, royal jelly is tricky to harvest, disturbs the hive and requires specialised knowledge and equipment. We don't recommend attempting your own harvest, but if you're keen to try it, royal jelly is widely available from commercial producers. It can be taken in powder or capsule form or eaten fresh (although in that case, we think the sour flavour is very much an acquired taste).

Propolis is an antimicrobial substance that bees use to seal up gaps in the hive and help protect the colony from pathogens.

HONEY RECIPES

Dairy-free Honey Banana Loaf

MAKES 1 SMALL LOAF

INGREDIENTS

½ cup (125 ml) extra virgin olive oil

½ cup (175 g) honey

2 eggs

5 bananas, 4 mashed (about 2 cups), 1 sliced lengthways

1 teaspoon baking soda

1 teaspoon sea salt flakes (kosher salt)

1 teaspoon cinnamon

1 teaspoon vanilla extract

1 cup (150 g) plain (all-purpose) flour

¾ cup (110 g) wholemeal (whole wheat) flour

vegan butter, to serve

This is an excellent recipe for using up all your ripe or overripe bananas! Some things to note: replacing sugar with honey in baked goods can make your mix wetter, so we've added a touch more of the dry ingredients here to balance. Honey will also brown more than sugar when cooked, so to prevent burning, we cook this on a low heat for slightly longer than usual. Using a glass loaf pan instead of a metal one will also reduce the browning.

Preheat the oven to 160°C (325°F). Use about 2 tablespoons of the olive oil to generously grease the base and sides of a loaf pan; it's okay if some pools at the bottom.

In the bowl of an electric mixer, mix the remaining olive oil with the honey for 1 minute. Add the eggs and mix for a further minute, then fold in the mashed banana with a spatula until combined.

Fold in the baking soda, salt, cinnamon and vanilla extract, then fold in the flours until just incorporated and no dry spots remain.

Pour the batter into the pan and top with sliced banana. Bake for 1 hour, then insert a skewer into the centre of the loaf; if it comes out clean, it's done. If not, cook for a further 5–10 minutes or until cooked through.

Allow to cool in the pan for at least 10 minutes before turning out. The loaf should pop right out if you flip the pan upside down, but if you need to give it a boost, you can run a knife along the sides before flipping it over.

Best served warm, with vegan butter spread on top for the butter lovers.

Roasted tomato sauce to use in everything

MAKES 2–3 CUPS (700 G)

INGREDIENTS

1.5 kg (3 lb 5 oz) medium–large tomatoes, halved

1 small brown onion, skin on, halved, or 1 leek, trimmed and cut into rounds

6 garlic cloves, skin on

3 tablespoons olive oil

1 tablespoon chopped fresh oregano or 1 teaspoon dried

Good handful of fresh basil, torn or chopped

1 tablespoon honey

½ teaspoon salt (or to taste)

Pepper to taste

An age-old method of reducing the acidity of tomatoes in sauces is to add a little sugar or balsamic vinegar. Honey also does this job beautifully, as well as adding a whisper of sweetness to the complexity of the flavours. You could even add a little extra to entice kids to gobble up more veggies.

Use this sauce as a base in pasta sauces, lasagne, meatballs or Moroccan dishes, or straight up with eggs, roasted vegetables, grilled meats or fish.

Preheat the oven to 200°C (400°F) and line a large roasting tin with baking paper.

In a large bowl, gently coat the halved tomatoes, onion or leek and garlic with the olive oil.

Place the tomatoes cut-side down in the tin, and scatter the onion and garlic around the tomato. Roast for 30 minutes, then add the oregano and three-quarters of the basil and roast for 10 minutes more or until skins are peeling off the tomatoes. Remove from the oven and set aside to cool.

Once cooled, pick out the onion and garlic skins and discard (and the tomato skins if you wish), then pour everything, including the roasting juices, into a food processor. Add the honey, then process until smooth. Season with salt and pepper, then taste and adjust flavour if needed.

Reheat the sauce if using straight away, and garnish with the remaining basil to serve.

If making for later, transfer to an airtight container and keep in the fridge for up to 5 days, or freeze for up to 6 months.

Honey-roasted macadamia nuts

MAKES 1 CUP (155 G)

INGREDIENTS

¼ cup (90 g) honey

Pinch of salt

1 cup (155 g) raw
    macadamia nuts

These simple but outrageously moreish nuts will find their way into many dishes. We use them in green salads, as a topping for desserts and in granola mix, and they are absolutely perfect for a cheese board. Be warned: if left unattended, they will disappear by the handful before they make it into any dish.

Preheat the oven to 180°C (350°F) and line a baking tray with baking paper.

In a bowl, mix the honey and salt with the macadamia nuts, then pour onto the lined tray.

Bake for 5–8 minutes, until the honey is bubbling and the nuts turn a light golden brown. The honey will look runny, but don't worry; the sugar in the honey will crystallise and harden slightly as it cools.

Set aside to cool for at least 10 minutes before eating. Store in a sealed jar.

Honey-roasted peaches

Peaches (or nectarines)

Honey

This is a very simple dish that is perfection all by itself, although you could also add a little olive oil and fresh thyme to the baking tray. It makes a perfect dessert served with cream or ice-cream, or breakfast served with yoghurt, granola, pancakes or waffles. It is also magic served with goat's cheese on a gourmet cheese board.

Preheat the oven to 220°C (425°F) and line a baking tray with baking paper.

Slice the peaches in half and remove the stones. Place them cut-side up on the lined baking tray.

Drizzle with honey and bake for about 20–25 minutes or until soft, depending on the size and ripeness of your peaches.

Chilli bin jam

MAKES 1 CUP (250 ML)

<u>INGREDIENTS</u>

250 g fresh red chillies

1 cup (250 ml) rice vinegar
or white wine vinegar

½ cup (175 g) honey

Our office manager Trace's chilli bin jam is a staple at our lunch table (the name is a play on the New Zealand term for an esky or cooler). This jam literally goes on anything and everything, and is a great way to use up the chillies in our Flow Hive garden.

Roughly chop the chillies and deseed according to the heat level of your chillies and how spicy you like your jam (more seeds equals more heat!).

Add the chillies, vinegar and honey to a small saucepan. Bring to a boil with the lid on, then reduce the heat to low and simmer for 15–20 minutes. Remove the lid and simmer for 10 more minutes.

Blend with a stick blender, then tip into a jar immediately, before it cools and thickens.

This will keep in the refrigerator for up to 6 months.

Honey mustard dressing

**MAKES ENOUGH FOR
A SALAD FOR 4 PEOPLE**

INGREDIENTS

3 tablespoons Dijon mustard

3 tablespoons extra
virgin olive oil

3 tablespoons apple cider
vinegar or fresh lemon juice

1 tablespoon honey

1 small garlic clove, minced

a generous shake of salt
and pepper

Put all the ingredients into a small jar with a lid, or in
a small bowl.

Pop the lid on and shake the jar to combine, or whisk
together in the bowl.

Best prepared right before using.

Balsamic secret dressing

**MAKES ENOUGH FOR
A SALAD FOR 4 PEOPLE**

INGREDIENTS

3 tablespoons extra virgin
olive oil

3 tablespoons balsamic vinegar

1 teaspoon honey

1 teaspoon kecap manis (or
sweetened soy sauce)

1 small garlic clove, minced

a generous shake of salt
and pepper

Put all the ingredients into a small jar with a lid, or in
a small bowl.

Pop the lid on and shake the jar to combine, or whisk
together in the bowl.

Best prepared right before using.

Chopped cucumber, snowpea sprouts, green leaves and flat-leaf parsley with honey-roasted macadamias (page 187) and honey mustard dressing.

Scan here for videos relating to this chapter

7

The
beekeeping
year

In beekeeping, one of the richest and most important areas of learning is the effect of nature's cyclical changes on what happens inside the hive. Since so many different aspects of the local ecology combine to influence a bee colony's health, beekeepers tend to become attuned to yearly cycles as they observe the constant subtle changes in the environment.

Gaining a stronger understanding of seasonality in your region will mean that, as well as being able to give your bees better care, you'll have a way to connect with nature very directly. You might find yourself starting to wonder things like 'What's in flower right now?'; 'Is that happening earlier or later than it did last year?'; 'How cool is the winter going to be, and for how long?'

In this way, beekeeping puts us in immediate relationship with the natural world, helping us to engage with our sense of place and with age-old cycles of renewal.

As honey bees have adapted to live in varying conditions worldwide, the specific way that your bees' needs vary throughout the year will depend on where you live; for example, bees may be active all year in warmer places and will possibly require care during periods of excess heat or humidity, while in freezing climates they'll bunker down over winter and only emerge from the hive again once the days begin to get warmer.

Climate can be very location specific. Here in the Northern Rivers, Australia, we can often harvest all through winter, whereas if you drive just two hours inland there'll be no honey at all. As far as bees are concerned, it's all about the presence of nectar and pollen (i.e. flowers). We recommend sourcing bees from quality breeders in your region as they are already adapted to your specific conditions. We also recommend learning from other beekeepers about what to expect in your area.

Temperate climates experience the full four seasons – spring, summer, autumn and winter – although the length and strength of each will vary according to your latitude and local conditions. In the Northern Rivers we are on the border of subtropical and temperate climates, so we get a bit of both. You can't often go by the seasons for what will be in bloom here, because Australian flora often responds to circumstances more than the time of year. Eucalypts flower irregularly, for example, just as kangaroos can breed at any time.

In Australia, First Nations peoples recognise many more distinct seasons than just four, based on closely observing local plants, animals and weather patterns. Their seasonal calendars reflect deep ecological knowledge and cultural connection to country that has been passed down through many generations. For the purposes of this chapter, we'll refer to the four seasons that are widely known internationally.

Spring

As the Earth wakes up after winter and flowers start to bloom, nectar and pollen are to be found in abundance. Bees use this supply of energy and nutrients to increase their numbers by building comb and raising new generations of workers and drones. It's often the easiest time of year to get bees, and usually provides great conditions for them to thrive in their new home.

This is the busiest and most exciting time in the beekeeper's year! As the days get longer and warmer, the queen ramps up egg production to increase the colony's size. Protein-rich pollen is abundant and the bees use it to make their bread and feed it to the developing brood. This is the time to do regular brood-box inspections. By mid-spring you could be inspecting your hive as often as every fortnight to monitor its health, make sure any unwelcome guests such as varroa mites aren't getting out of hand, and look out for signs of swarming.

Spring is also a great time to start a second hive. There are lots of benefits to having multiple hives. It allows you to share resources between colonies if one of them is struggling, and it makes it much easier to requeen if needed (see page 227 for instructions). You'll also learn more quickly when you can compare what's happening in one hive to another. Plus, of course, you'll get extra honey!

Swarm season

Have you noticed any new queen cells? These are specially adapted peanut-shaped cells about 3 centimetres (1 inch) long that protrude from the comb surface. They may be a clue that the hive is preparing to swarm, as when swarming behaviour is triggered, workers will make several queen cells. Shortly before the new queens emerge, the original queen will leave the hive with half the bees.

Swarming is a natural behaviour; it's the way that colonies reproduce. Along with producing drones, swarming is how each colony perpetuates its DNA. Although it's natural for the bees to swarm, it weakens your colony, can concern your neighbours and often results in more feral honey bees, so it's best to try to prevent it.

There are conditions that prompt the swarming instinct and signs that a hive is actually preparing to swarm. These conditions include plenty of food (nectar and pollen), the hive crowded with bees, and little space to store honey or lay more brood. When the hive is crowded you will see the bees 'bearding' – clustering around and below the entrance of the hive. They are out there to make room inside the hive for ventilation and so the rest of the workers can move about more easily.

The signs you should look for in your regular inspections (particularly in the spring) are the presence of queen cells at the bottom of frames and limited storage space. Some colonies, and some breeds of queens, will swarm more readily than others. If you have a hive that seems to swarm too easily or does so multiple times, you should probably requeen the hive with a queen from a reputable supplier.

It's possible to reduce the likelihood of swarming by providing more space in the hive. You might harvest some honey, add an extra brood box or a super, or remove a few honey-filled brood frames from the brood box and replace them with empty frames. Our go-to method of swarm prevention is to split the overcrowded hives, especially in spring (see page 202).

There's info on how to catch a swarm on page 103.

In spring, check hives as often as fortnightly.
Following pages When building new comb, bees form chains by hanging on to each other's legs. This is called 'festooning'.

Splitting a hive

In beekeeping, splitting means turning one hive into two – taking half the resources from one hive and putting them into another. This is an excellent way to both grow your apiary and help prevent swarming. If you don't want another colony, someone else surely will. Although hives are usually split in spring, as explained on page 198, the best time of year will, of course, depend on what's happening with the ecology in your area.

How to split a hive

Your aim is to end up with two healthy hives, each with a laying queen. There are various ways of achieving this but we will describe one of the most common and simple methods here. Before you start, you'll need to consider the following things:

- You'll need another brood box, with a base and a roof, and a set of assembled brood frames.

- One of the hives is going to need a new queen. You can choose between letting the bees make their own queen and buying a mated queen from a breeder and installing her.

- Do you want to split the hive evenly or just take a small split? If you want to reduce the chance of swarming without weakening the 'parent' colony, you can take just two or three frames for the 'child' colony.

- The post-split placement of the hives will be important. You can take one away to a different location, but it will end up weaker than the other as any foragers who were out while you made the split will return to the original hive. It's usually best to have the hive with the new queen in the original site and put the hive with the old queen less than a metre away. This will help ensure healthy population numbers in both hives. A few days after the split, if you find that one colony is weaker, you can place the weaker colony in the site of the original hive so that returning foragers will help boost its numbers.

- If you don't have much space you can go vertical, placing one hive on top of the other. It's also possible to split within a hive using a divider board.

So you're suited up with a lit smoker in hand, ready to get started. Place the empty brood box on top of its base beside the hive you want to split. Open the hive, remove up to half of the brood frames from the original hive and put them in the centre of the empty brood box. If you're making an even split you should try to divide the resources equally between the two hives. Both need bees, eggs, larvae, honey and pollen. Once you've transferred the frames, fill the remaining space in both brood boxes with empty frames.

If you decided to buy a queen, you'll first need to find the existing queen so that you know which hive needs to be given the new one. After splitting the hives, wait at least eight hours before adding the new queen, otherwise the chance that she could be rejected and killed is heightened. There's still a chance of this happening after eight hours – sometimes bees just don't accept a new monarch. Breeders supply queens in a small cage that has an entrance blocked by a candy plug. She will be accompanied by some escort bees. Place the queen cage between two central brood frames, with the cap on the upper side (so that if one of the escort bees dies, the exit won't get blocked). Tilt the exit slightly downwards so that if the candy melts on a hot day it won't drip on her highness. The workers will chew through the candy to release the queen, by which time the colony should be accustomed to her pheromones.

If you'd prefer to let your bees raise their own queen, you don't necessarily need to know which hive the original queen is in. If you can't find her when you're splitting the colony, just make sure that both hives end up with at least one frame containing eggs. Inspect both colonies four to five days after splitting. You'll know which hive has a queen (described as being 'queenright') when you find new eggs. Any eggs that were in the cells at the time of the split will have hatched into larvae by now. The colony that doesn't have eggs should have several well-developed queen cells by this time, so both hives can now be left to carry on. Check the hive with queen cells in another two-and-a-half to three weeks. By then a new queen should have emerged, mated and started laying eggs. If you have trouble finding her, just look for eggs. If you still don't see eggs a month after the hives were split, you should provide another frame containing eggs from your first colony or buy and install a queen from a breeder (or both).

Move frames from one brood box to another
to split a hive.

Dead outs

If you find that your hive has no brood and next to no bees, we call this a dead out. (Although springtime might be the most common time of year to discover that a hive has died, it's possible in any season.) This can happen for a number of reasons: perhaps your bees didn't make it through a long cold winter, or maybe they became queenless and didn't raise a new queen so their population slowly dwindled away, or maybe there was a pest or disease issue.

Regardless of the reason, a dead out should be removed from the apiary as soon as possible to avoid a situation where bees from other hives rob the remaining resources and potentially spread pathogens to their colonies. Before reusing any equipment, carry out an inspection to ascertain whether a brood disease such as AFB (page 232) or EFB (page 233) is present. If you're not sure what killed your bees, we don't recommend reusing any remaining comb in another hive.

Another reason dead outs should be removed promptly is that if bees get a taste for robbing honey from a dead hive, they can 'break bad' and start robbing other weak colonies rather than foraging.

Honey-bound hives

Honey-bound is a situation when there's so much honey stored in the hive that there's nowhere for the queen to lay eggs. This can trigger the hive to swarm, especially if it's springtime. For this reason, it's a good idea to harvest excess honey in spring, allowing the bees to transfer their honey stores from the brood box to your honey super and making room for the queen to lay her eggs.

If the season is well underway and your bees have full frames of honey on the edges of the brood box, it may pay to remove some of it. If you're using foundationless frames, you can cut out the comb and put the empty frame straight back in the hive towards the centre of the brood box. If you're using foundation, you should replace it with an empty frame.

This is an important part of spring management – it's always a balance between keeping your hives strong with the right numbers of bees and mitigating the risk of swarms.

Varroa management

Spring is an important window for varroa mite management. As colonies are building up, brood production increases and mite populations can explode if left unchecked. Staying ahead of the curve in spring will give you healthier hives and better honey yields later. We've got useful info on monitoring and treatment options for varroa on page 244.

Cycling out old brood frames

Over time, the wax in your brood frames will become darker. The cell walls will grow thicker as the bees re-wax them repeatedly, giving the pupae less space to grow and resulting in smaller bees. Residues and contaminants can also build up in the wax as time goes by.

When you see that the comb in your brood box is turning a really dark brown colour, it's time to start letting the bees build afresh. Spring is a good time for this as they will have lots of resources for building new comb. Most beekeepers do this by taking two or even three frames of old comb out of the brood box (avoid removing frames with brood in them) and inserting the same number of empty frames in the centre of the box. The following year you can repeat the process.

If you are using foundationless frames, you can simply cut out any old brood comb that doesn't contain brood, typically the edge frames, and put the now empty frames straight back in the hive for the bees to build afresh. Putting the frames back in towards the centre of the brood nest is a good idea, as the bees work from there outwards and will more quickly draw a fresh new comb for the queen to lay in. When discarding the old comb, make sure bees can't get access to it to limit the spread of pathogens.

If you find that an edge frame contains a small amount of brood but you wish to cycle the old comb out, you can either cut the bulk of the old wax comb out, leaving the section with brood, or you could utilise the roof cavity of the gabled roof. To do this you take out the plug in the inner cover and place the frame on its side. Make sure it's propped up with at least a 5 millimetre (³⁄₁₆ inch) gap underneath so the bees can service the comb. The nurse bees will then come up into the roof cavity and nurture the brood until it emerges. After that the frame can be removed.

Old comb becomes dark in colour and should be cycled out of a hive.

False springs and supplemental feeding

As a hive builds up its numbers in early spring in anticipation of an influx of nectar and pollen, the colony can be vulnerable if it's prevented from foraging by a prolonged cold snap or rainy period. If this happens, you can easily rescue it by feeding the hive. Some beekeepers feed their bees pollen supplements during springtime or times of dearth to provide them with extra protein for raising brood. (See more on feeding below.) Giving bees probiotics to improve their gut health and immunity is a relatively new practice that's beginning to grow in popularity.

Nutrition deep dive

The main thing you need to know about bee nutrition is that the colony needs carbs, protein and water to be healthy. If they haven't got enough of any of the above, you may need to give them a helping hand. Here are more details for those who are keen to gain a deeper understanding.

Just like humans, bees require a varied and balanced intake of water, protein, carbohydrates, vitamins, minerals and lipids (fats). Workers, drones and queens require different diets, but they all get just about everything they need from nectar, pollen and water. Their carbohydrates come in the form of nectar, which consists primarily of sucrose, glucose, fructose and water. As they process the nectar, they add an enzyme that breaks down the sucrose.

Pollen, bees' main source of protein, is also rich in vitamins, minerals and lipids. They typically prefer to collect pollen from a diverse range of plants to ensure that they get all of the ten amino acids that are essential to their wellbeing. Crude protein percentage (CP%) is a measurement of the amount of protein present in a substance. Bees prefer a CP% of 23 to 30. Different plant species produce pollen with varying CP%; for example, sunflowers have around 15 per cent and almonds 26 per cent. The higher the CP% of your local flora, the less the bees will need to forage to sustain the hive.

Insufficient protein in a colony's diet may result in shorter lifespans, weaker immunity, reduced population and extended time for drones to reach maturity. While you can't easily measure pollen availability or quality, there are signs that will tell you if your hive has what it needs. If you see foragers returning to the hive with varying colours of pollen, or if you see bee bread in multiple colours, this means they are collecting pollen from multiple sources.

It's a good sign if you see larvae 'swimming' in royal jelly. If they look dry, then you may need to consider supplementary feeding.

Supplementary feeding

When the local environment doesn't naturally have enough of what your bees need, you can use supplements to get them through. Supplementary feeding can stimulate bees to rear more brood or to forage for nectar and pollen, can provide the colony with stores for lean periods and can promote hygienic behaviour.

Carbohydrate supplements (sugar)

White sugar (sucrose) is the most common nectar/carbohydrate supplement used by beekeepers. Other sugars don't have the same effect, and often are not as easy for the bees to digest. If there's a long nectar dearth, or if you're going into winter and your bees don't have enough stores, it's best to feed them a thicker syrup (two parts sugar to one part water) that they will store for later. If you feed with a thinner syrup (often 1:1 ratio), it's more likely to stimulate brood rearing and foraging. You can also feed bees dry sugar; however, this is usually only done in cold or emergency situations as they then need to use water to dissolve it.

Making a sugar feed

Mix sugar with warm water to make a syrup and stir until all the crystals have dissolved. When it cools down, pour it into a clean jar, then punch a dozen tiny holes in the lid. Upend the jar feeder on your hive's inner cover with the central plug removed, so the tiny holes in the lid are over the hole. Another method is to pour the syrup into a resealable plastic bag, squeezing it to expel any air before pricking tiny holes in the bag. Put the bag feeder under the roof, sitting on the inner cover with the plug removed, so the bees can find it if they need extra food. Make sure the feeder is inaccessible to other hives. If you'd like to make your own bee feeder, see the link to the DIY demonstration in the Resources section on page 276.

Protein supplements

There are many pollen/protein supplements available on the market, usually in dry form or as wet 'patties', which are a mix of pollen, soy flour, yeast, sugar and water. Note that pollen feeding in the wrong circumstances can induce premature swarming, and sometimes bees will consume patties not because they need the protein but rather to extract the sugar – so try to be aware of what your bees actually need.

If you're feeding a patty to an already weak colony, the bees may become more susceptible to wax moth and small hive beetles. Ensure you monitor frequently and don't supply them with more than they can manage.

Pollen supplements should be stored in the fridge.

Where the supplement is placed in the hive can affect uptake. Bees are often more likely to consume a protein patty the closer to the brood box it is located.

How much should I feed?

The amount of supplementation necessary will depend on the colony and when these resources will next be available naturally.

Bees generally prefer to consume what they can forage. If your bees stop consuming the supplement, conduct an inspection to see if they are bringing enough in naturally.

Types of feeders

Feeders come in many forms: some fit inside the hive roof; some are small boxes that can be placed on top of the brood box or super; frame feeders hang inside the box; and there are also entrance feeders.

Open feeding is also done, but it's not considered best practice because of the risks of spreading pests and diseases and the fact that stronger colonies can crowd out weaker ones. In some jurisdictions it is prohibited because of biosecurity risks.

A tray feeder in the roof cavity of a Flow Hive.

Summer

The nectar available for bees to collect at any given time of year – also known as the nectar flow or honey flow – depends on which plants are in flower, along with climatic factors. Nectar flow should continue throughout summer and may even increase. Keep up regular hive inspections and varroa mite counts, but it's usually okay to do them a little less often than in spring. Swarming may also continue or ramp up for a second time. If your hive is going strong and you haven't yet added the super, now is likely to be a good time to do so (see instructions on page 151).

Beating the heat

If you live in an area where the summers are sweltering, you may want to position your hive so it receives afternoon shade. Excessive fanning and a lot of bees hanging on the front of the hive is a sign that they may be heat-stressed. Consider making adjustments to allow more ventilation or removing the bottom tray to get more airflow into the hive (remember to replace the tray when temperatures drop). Bees are very good at managing their own air conditioning, so you might not need to do anything. Hive ventilation is a topic where opinions among beekeepers tend to vary greatly! If you live in a wet climate, it's generally recommended that you provide enough airflow to prevent excessive moisture building up. When excessively dry weather occurs, water supply is crucial. Setting up a water station can assist your bees (for more info see the Resources section on page 276).

Above Bees need access to a supply of fresh water. **Right** Bearding is a way bees regulate the hive's temperature during hot summer weather.

Boozy bees

Under certain conditions, nectar can ferment while it's still in the flower. Normally the high levels of sugar in nectar prevent microbial growth, but if high rainfall and/or humidity dilute it during hot weather, then natural yeasts can cause fermentation to begin.

Researchers have found that although most animals don't willingly drink alcohol, honey bees actually seem to like it! Relative to their body size, they'll willingly drink the equivalent of a human downing 10 litres of wine in one sitting. They'll even drink pure ethanol and will choose to drink from an artificial flower containing 5 per cent ethanol over one containing only a sucrose solution.

Bees that forage on alcoholic nectar (or discarded wine bottles, which does happen) usually can't find their way back to the hive; they fly into things or just fall to the ground. Those that do make it back are prevented from entering the hive by guard bees (or maybe they should be called bouncer bees in this case!) so that the fermented payload they're carrying doesn't put the entire colony at risk.

Stormy weather

Where we live, summer can bring lots of storms. If strong winds are headed your way, there are some precautions you can take to ensure your beehives will be okay. Firstly, your hive should be stable and on a solid footing, as a downpour combined with high winds could reduce the earth below your hive to mud, making the hive more susceptible to toppling. To prevent this, something solid such as a paver or a wood block can be placed under each corner to stabilise it.

If you've recently added a super to your brood box, your bees may not yet have had a chance to fully glue the boxes together with propolis. In that case, you could wrap a strap around your hive from top to bottom to make sure the super won't blow off. If your hive is more well-established with a relatively full super, you may not need to do anything. A half-full super is quite heavy and once it's been glued in place with propolis it can handle very strong winds.

Newer Flow Hive models have roof locks attaching the roof to the hive, but if yours doesn't have these you can simply strap the roof firmly to the rest of the hive so it doesn't blow off. Be careful not to use too much pressure or you might snap the roof shingles.

Autumn (Fall)

NOTE: If you live in a warm climate where bees can forage throughout the winter season, then most of what we've written in this section won't apply to you.

Consider your bees' ongoing access to pollen and nectar in late summer and early autumn. Be aware of the honey levels in the hive. The bees will eat this honey throughout winter, and you should leave enough to get them through the colder months. The right time to stop harvesting may change from year to year. You can seek advice from other beekeepers if you're unsure what to expect in terms of nectar flows.

As the days become shorter and the weather cools, the queen produces fewer eggs, reducing the size of the brood nest and the overall number of bees. The colony will raise 'fat bees' in preparation for winter. These are a special type of worker bee that can live for several months without foraging. Compared to regular workers, they can store more nutrients in their bodies and they also have stronger immune systems.

Before winter starts you should make your last thorough inspection of the hive to see how much honey they've put away and to check for pests and illness. It's essential to deal with varroa mite infestations above a certain threshold at any time of year, but bees are particularly vulnerable during winter. If mite levels are high, be sure to take action (see page 244 for advice on what to do).

Depending on the amount of honey in the hive, you might need to add a sugar feed before winter. It's not good to open the hive during winter when it's very cold, so now is the time to make sure your bees have the supplies they'll need.

A bee drinking sugar water from a feeder.

Preparing for winter

As the end of autumn approaches, in colder climates you may wish to pack down the hive. This means reducing the size of the hive so that the space the bees need to keep warm becomes smaller. As with most things in beekeeping, there are multiple methods available and people have different preferences. According to how long and cold the winter is expected to be, and how big your colony is, you might pack down to just a brood box or two brood boxes, or a brood box plus an ideal (shallow) honey super. An experienced beekeeper in your area should be able to give you some tips.

If you've decided to take your super off for the winter, storing the Flow Frames in a freezer is a good way to keep any remaining honey or nectar still in the frames from fermenting. If you've decided to harvest before taking your super off, you can leave it on for a couple of days after the harvest to let the bees clean it up. If you can't store your super in the freezer over winter we recommend washing your Flow Frames in warm water to remove any honey. It's best to set the Flow Frame in the harvesting position so the water can easily wash through the cells and out the trough. The frames should be air-dried thoroughly in the shade before being put away in a dry, dark place. Clean and store the plastic queen excluder, too.

Throughout a cold winter, the bees will stay in a tight cluster and consume their honey stores. If you're not packing down and have left food stores in a super above the brood box, it's important to remove the queen excluder before winter so that the queen can climb up with the rest of the cluster. Otherwise she'll have no food and may perish in the cold.

If it gets particularly cold in your area, you might consider adding insulation to the hive's roof cavity. Some beekeepers also create a windbreak (for example using hay bales) or use insulative wraps around the hive body. Check to make sure your hive's roof is on correctly, and that there are no gaps between boxes that will let in draughts. Of course, the bees still need fresh air, but the ventilation at the base of the hive should be enough to provide oxygen and expel carbon dioxide. An entrance reducer can be handy for reducing draughts, but it can become blocked by debris so may need checking periodically.

Winter

Throughout the colder seasons, as long as there's no forage available and/or it's too cold to fly, honey bees stay inside the hive to keep warm. Once you've prepared the hive as per the instructions in this chapter and they're comfortably settled, you won't need to do much else for them, provided the colony is healthy and has enough reserves of honey or feed to see them through to early spring.

Worker bees will often expel drones from the hive at the end of autumn or the beginning of winter, dragging them out and sometimes even biting off their wings. Because the large drones consume a lot of precious resources, and there's not going to be any mating happening in winter, they're considered expendable. This is bad news for drones, but it's how bees have evolved to give the colony its best chance of surviving until spring.

Hunkering down

At this time of year, the workers cluster tightly around the queen at the top of the hive. As the temperature drops, to keep the hive at an optimum temperature of 34–36 degrees Celsius (94–97 degrees Fahrenheit), the worker bees detach their wings from their wing muscles and vibrate their bodies to create heat. The queen stays at the centre of the cluster, while the workers take turns rotating around the edges to make sure that no one bee gets too cold.

Don't open the hive when it's very cold as you'll let the heat out! Generally you shouldn't open your hive if the temperature is below 15 degrees Celsius (59 degrees Fahrenheit); however, if you're in a cooler climate and the bees are well adapted, a very brief inspection may be possible above 10 degrees Celsius (50 degrees Fahrenheit) as long as you take care not to allow the brood to get too cold. Once the bees have stopped flying for the winter, it's best not to disturb them until early spring, when you will notice bees buzzing around the entrance to the hive as the weather warms up. They'll start to take cleansing flights (i.e. bathroom breaks) once the weather is fair enough.

Ice and snow

Snow acts as an insulator around the hive and doesn't necessarily need to be cleared away. After a deep snowfall, you might see little tunnels around the hive which the bees have made for ventilation and so that they can go out to collect water.

Seasonality in the tropics

Beekeepers in temperate climates are always conscious of the season, aware of their bees' foraging opportunities in the spring and summer, and making sure they have enough honey stores to see them through the winter. In tropical climates, honey bees don't have a 'winter' season. They can usually access food year-round. However, tropics-based bees have other challenges that need attending to. During the wet season and droughts, access to forage can be affected so it's still a good idea to keep an eye on honey stores in case you need to feed a hive that's out of resources.

Hot, humid conditions can lead to pests, such as the small hive beetle (see page 240), increasing in numbers, while chalkbrood (page 234) thrives in damp conditions, so these issues may need attending to. Although seasonal demands in the tropics may differ from those in the rest of the world, most of the same beekeeping practices are relevant.

MEET THE BEEKEEPER

A beekeeping romance

Victoria, Alaska, USA

Victoria is a grandmother, beekeeper and pilot. After falling in love with a beekeeper, she soon fell in love with beekeeping too. To see more of Victoria's story, check out the Resources section on page 276.

'I fell in love and he had bees. I would stay inside the house and watch. It took me a year to get comfortable, then I wanted a Flow Hive. Now you can't keep me away from them.'

MEET THE BEEKEEPER

Boosting the garden with bees

Kiana, Washington, USA

Firefighter Kiana has been amazed by how much her home produce has flourished since she added Flow Hives to her backyard. To see more of Kiana's story, check out the Resources section on page 276.

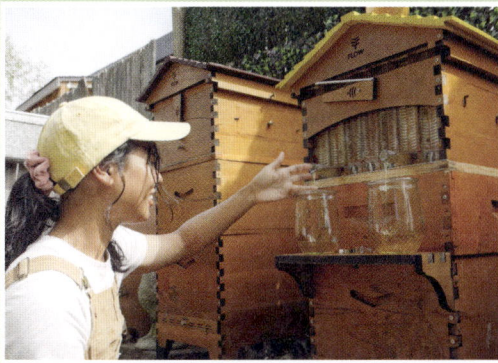

'It boosts pollination, so everything does better. My garden just blows up during the summertime.'

Scan here for videos relating to this chapter

8

Hive health
and
troubleshooting

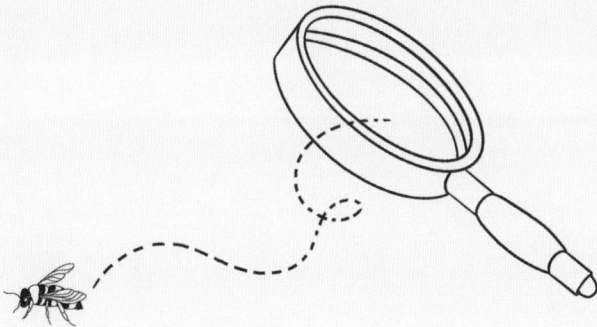

This guide tackles many of the most common issues beekeepers will encounter. If you don't find the answer you're looking for below, check out the Resources section on page 276 for links to our detailed online information.

Bee behaviour outside the hive

This section covers what actions to take if you notice your bees doing unusual things.

Bees clustering outside the hive, AKA 'bearding'

Bearding is when bees form in a clump outside the hive, often hanging in a shape resembling a beard. This is a natural occurrence and helps cool and ventilate the hive in very hot weather. However, bearding can also be a sign of overcrowding, which could lead to swarming. If you look in the observation windows and can't see the comb because there are too many bees, it's time to do an inspection and be prepared to implement swarm prevention measures (see page 198).

Low activity at the hive entrance

A regular glance at the hive entrance is often useful to gauge how your bees are going. A lack of activity at times when your bees would normally be flying can be a cause for concern, or it might be that the weather is unfavourable.

A quiet landing board might indicate that the colony has become extremely weak or died out. If it's not too chilly (below 15 degrees Celsius or 59 degrees Fahrenheit) you should perform an inspection to see what's going on (see page 130).

Angry/aggressive bees

Weather can influence your bees' temperament. If possible, avoid opening the hive in cold or windy weather; wait for a fine, still day. Make sure you move gently and use more smoke than usual if necessary. Some floral sources impact the bees' temperaments; for example, canola is said to make them crankier.

If your bees are consistently aggressive – many beekeepers use a 'three strikes' rule – it's quite likely the queen's genes are influencing this behaviour. It is difficult dealing with cranky bees every time you open the hive, and of course they present more of a risk to people close by. In these circumstances, it's common for beekeepers to remove the queen and replace her with a more docile breed. See the notes on requeening on page 227.

Bees swarming

It's normal to see lots of activity at the front of the hive as young bees take their orientation flights, but this is often confused with swarming behaviour. When bees are swarming you will see a clear flow of marching bees exiting the hive and joining an ever-growing cloud of bees in the air. Sometimes they'll do this and return to the hive, but other times half the bees will leave permanently. In this case they may temporarily land nearby (see the section on catching swarms on page 103).

If the swarm does return, you might want to take action straight away to split the hive, which will hopefully prevent the swarming behaviour (see page 202 for instructions on splitting).

Robber bees

Robbing is when bees from one hive invade another hive en masse to steal honey. It's characterised by a lot of frantic activity at the entrance of a hive, with bees aggressively darting in and out and sometimes fighting. It can be distinguished from regular activity by its chaotic rather than orderly nature. The main way to tell that robbing is underway is that robber bees will try to enter the hive through any crack or crevice rather than just the entrance. You may see dead bees on the landing board and inside the hive, and wax cappings on honey frames will be torn open.

Robbing must be stopped immediately as a hive can be robbed to death. If a sick hive is robbed the pathogens are likely to be spread to the robbers' hive. You can take the following steps to save your hive from robbers.

1   Do not leave sugar or honey exposed around any bee hive. When nectar is in short supply this can send a colony into robbing mode.

2   Reduce the size of the hive entrance to the width of a single bee so that it is easier for the guard bees to defend the hive.

3   Ensure there are no other points of access into the hive (check for gaps between boxes or under the roof – an easily identifiable sign is a group of bees actively trying to enter via a gap). Shove a small amount of dry grass or steel wool or a rag into the opening to quickly block it.

4   If the robbing continues, draping a wet bed sheet over the hive will really slow down and confuse any robber bees.

5   If the robbing is particularly frantic, you can close the entrance completely using some vegetation or a wet towel (but only for a short period). Make sure the hive has sufficient ventilation – use mesh to close the entrance or ventilate via the screened bottom board if you have one.

6   Some beekeepers like to use a sheet of glass or a robbing screen (a mesh box with small entrances at the top) placed in front of the hive entrance to confuse the robbers.

7   Once robbing has ceased, investigate the robbed hive to identify why the hive is weak, check for disease and/or queenlessness and take appropriate action.

8   Eliminate any extra space in weak hives by reducing box numbers or sizes to make them easier to defend. Feed the colony to strengthen if necessary (note that feeding can induce further robbing if not done carefully).

To prevent further attacks:

- Be aware of nectar availability and the strength of your colonies. Robbing is more likely during a dearth (lack of nectar). Be prepared to feed your bees during dearths (see page 207). In an apiary with multiple hives, it's often worth aiming to equalise the strength of your hives so that weaker colonies don't become vulnerable to robbing by the others.
- Do not feed honey or sugar water outside the hive, as this can encourage robbing behaviour.
- Reduce the entrance on weak hives and during times of low forage availability.

Brood issues

Healthy brood is the single most important sign that your colony is doing well.

No brood or eggs in the brood box

If you notice a lack of brood in your hive, it could mean a few things: your colony is queenless; the queen has stopped laying; the queen is newly mated and has yet to start laying; or you have a virgin queen.

If you suspect that you have a virgin queen, check the brood box again in two weeks. If there are still no eggs or brood, requeen the hive (see page 227) or add a frame with uncapped worker brood from another colony to the hive. The worker bees will build queen cells to raise a new queen (see our brood development calendar on page 67).

It is natural for brood rearing to stop in some districts that have a lengthy period of cold temperatures in late autumn and winter.

Drone brood in the Flow Frames or honey super

If there are no eggs or brood in the brood box, there's a chance the queen could be above the excluder in the super. If you discover worker brood in the super, find the queen and manually put her into the brood box. If there is brood in the super but you can't find the queen, brush or shake all the bees from the Flow Frames into the brood box. Check the queen excluder to find the reason the queen made it through and repair or replace it if necessary. When the queen is newly mated, she is sometimes small enough to squeeze through the excluder. Also check for gaps that the bees might be using as a top entrance above the excluder. A returning queen may have entered above the excluder if there is a top entrance.

After putting the queen excluder and super back on, prop open the inner cover and roof slightly with a small stick so that any drones emerging from the frames in the super may exit (they can't get back down to the hive entrance through the queen excluder). Inspect the brood box for eggs in two or three days to ensure that the queen is laying.

Another reason you might find brood in the Flow Frames is laying worker bees. If the hive is queenless, worker bees will start laying unfertilised drone eggs on cell walls all around the hive, often multiple eggs per cell. In this case, requeen the hive (see page 227) or add a frame with eggs and open brood so the colony can raise a new queen. To tell the difference between drone and worker brood, as it can be hard to tell in the deeper Flow Frame cells, take a look at the developing pupae. Drones have larger eyes that touch together, while workers' eyes are situated towards the sides of their heads.

Brood box contains only drone brood, no worker brood

This probably means that the colony is queenless and workers are laying unfertilised eggs that will become drones (see the next point below).

Another possibility is that the queen is poorly mated and cannot lay fertilised eggs, only unfertilised eggs. If the colony is not too weak and young bees are present, remove and replace the queen (see page 227). If the colony is weak, it may be bolstered with frames of brood from another hive.

Eggs laid on cell walls and/or multiple eggs laid in cells

Eggs on cell walls rather than at the bottom of cells can indicate that they were laid by workers with their shorter abdomens. If you see large numbers of eggs in single cells, then the colony has probably been queenless for a while and the workers have begun laying. These eggs are of course unfertilised and will hatch into drones. In this case you should requeen the hive (see page 227) or add a frame from another healthy hive with eggs and open brood so the colony can raise a new queen. However, a newly mated queen can often lay two or even three eggs at a time. So if you see multiple eggs in cells but worker brood is also present, it could be that you've got a newly mated queen and she'll settle into a better laying pattern soon enough.

Many eggs in cells are a sign that workers have begun laying in a queenless hive.

In this hive, a young queen has laid more than one egg in some of the cells.

Bald brood

Bald brood is when developing pupae are uncapped and continue to grow while exposed. This could be caused by wax moth larvae tunnelling through the comb and perforating cell caps in the brood box, or by instinctive hygienic behaviour by the bees as they uncap brood to deal with varroa mites, chalkbrood or foulbrood. If it's a minor case of bald brood and everything else looks okay, there's probably nothing to worry about. If there's lots of bald brood you might have an overabundance of wax moth or varroa that needs to be dealt with.

There could also be a genetic cause, and if this is the case requeening the hive should solve it (see page 227).

Scattered or irregular brood pattern

A healthy brood pattern indicates a strong colony. A patchy brood pattern could indicate the following.

- Brood disease: see pages 232–235 for detailed information on symptoms.
- Poor or failing queen: consider requeening with a newly mated queen.
- Starvation: if you see a lot of bees head-down in the cells looking for sustenance, it could be a sign that they're starving. Feed your bees or relocate them to an area with better conditions.
- Pesticide exposure: see page 230 for more. Remove dead bees and feed the colony with sugar water if required to build strength. If pesticide is likely to be resprayed, relocate the hive.
- Extreme weather conditions: see pages 210–215 for info on how to manage bees in very hot or cold weather.

Cap discolouration or unusual appearance

If you see sunken caps over brood cells, discoloured larvae in open or sealed cells or sunken dark caps with piercings in them, this is likely to indicate brood disease. Carefully inspect the hive. See pages 232–235 for information on pathogens.

Brood cells very dark in colour with thicker cell walls

Brood comb cells start light in colour and become darker the more they are used. It is common practice to cycle out older brood comb every three to four years so the bees can start afresh. See page 206 for instructions.

White to grey-black, chalk-like 'mummies' on hive bottom board and outside hive entrance (also in cells)

This indicates a condition called chalkbrood, which is caused by a common fungal disease. See page 234.

Queenlessness and requeening

Signs that your hive may have lost the queen include the following.

- No eggs or larvae in the brood nest.
- Supersedure or emergency queen cells (elongated peanut-shaped cells in the middle of a frame).
- Extra honey, nectar and pollen stores (this is because more worker bees will turn to foraging when there are no young to nurse).
- Drop in colony population.
- Eggs on cell walls instead of at the bottom of cells, multiple eggs in cells (see pages 224–225).

Requeening

You might introduce a new queen to your hive because the original has died or is unhealthy, or to introduce new genetics. A mated queen has enough sperm to last many years of laying. Her own DNA, combined with that of the drones she's mated with, will determine the genetics and temperament of the hive. Queens can be mail ordered from a breeder. She usually comes in a queen cage with a few attendant bees who will feed her while in transit. One end of the queen cage is blocked with candy; in the hive the bees chew through the candy over a few days to release the queen. This gives the colony time to become familiar with her new pheromone.

You can buy virgin or unmated queens, but the steps below describe the most common requeening method, which is to use a mated queen.

1   Once you have received your queen, keep her at room temperature and out of direct sunlight. Rub a little water onto the outside of the cage twice a day but don't wet the bees. Install the queen into your hive ASAP.

2   When replacing the queen, remove the original queen from the hive first, or the colony will not accept the new queen. Many beekeepers like to leave a gap of at least eight hours between removing the old queen and adding a new one.

3   Place the cage on top of the brood frames and assess the reaction of the hive. On rare occasions, they will crowd aggressively around the cage and try to sting her through it: do not introduce this queen yet. If the bees inspect the cage calmly and attempt to smell or feed the new queen, it's likely they'll accept her.

4   Place the queen cage into the hive between two brood frames, choosing a place where honeycomb is minimal so honey doesn't flood the cage. Place the candy plug on the upper side with the exit tube facing slightly downwards into the hive at an angle. This prevents the queen from getting trapped in the cage (for example, in case one of her attendants were to die and block the exit tube) and keeps any candy that may melt on a hot day from running back into the cage and onto the queen. (See the Resources section on page 276 for a video of this.)

5   Inspect the hive about a week after installing the queen. Check the brood cells for eggs, as these will show if the requeening has been successful. There is always a chance that the bees may not accept an introduced queen. If your requeening attempt hasn't been successful, you can try again or alternatively introduce a frame from another hive containing eggs and let the bees raise their own queen.

Lay the queen cage on top of the brood frames to see how the bees react.

Wedge the queen cage, with the exit tube pointing downwards, between two brood frames.

Bee loss

Colony absconded

If nearly all your bees are simply gone, it can mean that the colony has absconded. Bees may leave if conditions in the hive or in the environment are unfavourable; for example, an infestation of pests, harassment by predators, the hive falling over, extreme weather or the presence of toxic chemicals could cause the bees to leave.

If your bees have absconded, try to determine the factors that could have led to them leaving. Thoroughly clean the hive, checking it for signs of pests or disease. Repair structural issues if any are present. You can then repopulate the hive once conditions are favourable.

A large clump of dead bees lying head first in cells

Bees cluster head first in cells during cold winters. If they're dead and there are no honey stores, this indicates that the colony has probably starved. If there are still survivors, feed them immediately (see page 207).

Dead drones outside the hive entrance

It's common when forage is scarce for the worker bees to sacrifice drones. You may see this at the beginning of winter. If you watch closely, you may witness worker bees tearing at the legs and wings of drones and dragging them out of the hive to perish. Inspect the hive to ensure it has sufficient stores. If not, consider sugar feeding.

Dead or sick adult bees on the ground or on the bottom board

While it's normal for there to be a few dead bees near the hive or on the ground or bottom board, if there is an unusually large number then consider the following factors.

- There may be pesticides in the foraging range (see 230).
- If bees have deformed wings (K-wing), this could be associated with varroa mites (see page 244) or tropilaelaps mites (page 243), or diseases including Nosema (page 235).
- Starvation: check hive for adequate stores.
- Extreme heatwave: provide shade in summer, and make sure they have access to water.
- Fighting caused by robber bees: reduce the entrance width; once robbing has stopped, address underlying issues within the apiary (see the robber bees section on page 223).

If there are lots of dead bees around the hive, try to ascertain the cause.

Bee numbers are declining

If you notice that the number of bees in your hive has declined, it could be for the following reasons.

- If there is a lack of pollen and nectar, the queen may be naturally responding to the availability of forage and laying fewer eggs. If the numbers are declining and there are little or no honey stores, they may be starving. See the info on feeding on page 207.
- The colony has swarmed and half of the population may have gone off looking for a new home. This is the most common reason for a sudden drop in numbers. The bees should have raised a new queen and the population should build quickly again over the coming weeks. Check the hive for eggs and brood after two or three weeks. For more on swarming, see page 198.
- The hive has become queenless: requeen (see page 227), provide one or more frames with eggs and open brood from another colony or merge with another hive (see page 230).
- The queen is ageing, failing or infertile: requeen or merge the hive with a stronger colony.
- The population can dwindle after winter, especially in the colder regions. Feeding bees in early spring can help to boost numbers in a weak colony (see page 207).
- The colony has a disease that is slowing down reproduction: see the pathogens section on pages 231–235.
- High numbers of varroa mites: see page 244.
- There are ground-dwelling predators.
- Predatory wasps/yellow jackets prey on bees in some continents. Reduce the hive entrance so the colony can better defend itself, and destroy any wasp nests you can find in the vicinity.

- Harsh weather conditions such as extreme heat, drought, excess rainfall or prolonged cold can all impact population numbers.
- The bees have been poisoned by pesticides (see below).

How to unite/combine colonies

If a colony is weak, you may wish to combine it with another to save it. First, make sure that neither colony is diseased, and choose a period when there is nectar and pollen available and weather conditions are appropriate to carry out the merger. It's better to combine two weak colonies than a strong one and a weak one, to avoid the risk of health issues from the weak one spreading to the strong one.

If the weak colony has a queen, you will need to remove her. Next, open the other hive and place two sheets of newspaper over the brood box. Using your hive tool, make a few small slits or holes in the newspaper to allow ventilation. The newspaper will act as a barrier to slow down integration, allowing time for bees to get used to the new pheromones and preventing fighting. Place the brood box of the weaker colony on top of the other brood box. After a few days the bees will chew away the newspaper and the two colonies will become one larger hive.

Pesticides in foraging range

If many bees are found dead in and around the hive with their tongues sticking out, it's often a sign that they've been poisoned from pesticide exposure. Because bees forage in a wide radius, there is a lot of potential for them to be exposed to contaminants outside of the beekeeper's control.

If you know when pesticides will be sprayed in your area, you can temporarily close the hive entrance (ensuring adequate ventilation) to reduce exposure to chemicals. You can also consider moving the hive to a safer location. Document the incident and consider reporting it to your local agricultural or environmental body. If you can, try to identify the source of the event by speaking to neighbouring landowners to find out what was sprayed and when. (Note: In some regions there are online platforms that allow farmers to communicate with beekeepers about crop protection activities so that steps can be taken to minimise impacts.) Contact any farmers, companies or local authorities that are involved in spraying in your area (for example, mosquito control companies). Ask them not to spray near your hives, to spray at night or to inform you when they will be spraying so that you can take protective measures. Keep in mind that bees forage on flowers, so if the spray is on leaves, bark or lawns, it is unlikely your bees will be impacted. Some authorities will test bees for pesticide poisoning at no cost, while others will charge for this service. Speaking to neighbours and raising awareness of pesticides and their effect on bees is a great way to advocate for pollinators. A jar of honey is not only a sweet gift but a good conversation starter.

Pathogens

Monitoring your hive's health is an essential part of beekeeping. Look out for signs of illness, and, if necessary, seek advice from your local beekeeping association or authority to learn more about which monitoring, treatment, containment and prevention practices are most appropriate in your region.

Many beekeepers use a barrier system to increase hygiene and minimise the possibility of disease spreading from one colony to another. In this system, hive components are never swapped from hive to hive and gloves, tools and any other equipment are sterilised before being used with the next hive. If your hives are more than 10 metres (32 feet) apart this will really reduce the drift of bees from one hive to the next and reduce the risk of pathogens spreading. As 10 metres apart is often not practical, beekeepers place hives in groups to quarantine one apiary from the next.

Certain chemical and antibiotic treatments are restricted in some jurisdictions, so be sure to follow local regulations if applying any chemical treatments. Take care to sterilise equipment between inspections if working on multiple hives to reduce the probability of cross-infection. And if you buy any second-hand beekeeping equipment, you should make sure you sterilise it before use.

American foulbrood (AFB)

This is a highly infectious disease caused by the spore-forming bacteria *Paenibacillus larvae*. The spores are difficult to kill as they are resistant to extreme heat and can survive in a dormant state for 50 years or more. Spores are spread by bees robbing a weak hive, bees drifting between colonies, and unhygienic beekeeping practices. AFB usually results in the eventual death of the colony.

A rope of slime that will cling to a matchstick is a sign of AFB.

American foulbrood is a notifiable disease in many jurisdictions. Contact your local authorities immediately if you think AFB is present in your hive.

Symptoms: The first sign of the disease is an irregular brood pattern and sunken dark cappings, with small puncture holes in some of the cappings. You may also smell a foul odour in the hive.

If you see larvae or pupae with a thin thread (the remains of the pupal tongue) stretching across the cell from bottom to top, it's a sure sign of AFB.

Larvae that have died can dry out to form hard scales that stick to the bottom of the cells.

To get a better indication of whether AFB is present, inspect a perforated cell closely to make sure that the small punctures in the capping aren't from a new bee emerging, then poke a matchstick into the cell. If AFB is present, a thin ropy string of brown slime will cling to the cell and the matchstick, much like the way mozzarella will stretch from a slice of pizza.

Treatment: In many places, infected colonies must be euthanised and the hives burnt or irradiated to prevent the spread of the disease. Contact local authorities to determine how to proceed. See the Resources section on page 276 to read about Flow Frame sterilisation and irradiation.

Prevention: Employ strict hygiene measures such as maintaining separation of equipment between hives to reduce cross-infection, sterilising second-hand equipment, and not feeding bees honey from other hives. Some colonies show higher resistance to AFB through hygienic behaviour, a trait that can be selected by breeders. You may wish to purchase queens with hygienic traits from a breeder.

If you would like to get a diagnosis to confirm the presence of AFB or EFB (see opposite), many places in the world offer a service where you can send away a sample of the infected brood for testing. There are also infield testing kits available.

European foulbrood (EFB)

European foulbrood (EFB) is caused by the bacterium *Melissococcus pluton*. It affects bee larvae and can cause a significant reduction in the capacity of the hive. If left unchecked, it can lead to the death of the colony. Cool, wet weather and poor nutrition can make EFB symptoms worse.

Symptoms: Diseased larvae change colour from white to a yellowish brown; dead larvae may be watery; an uneven brood pattern with capped and uncapped cells mixed; sour odour.

Treatment: A brood break will generally fix EFB, and requeening is helpful. If you remove the old queen, it will give you both: the lack of a queen creates the brood break, and the bees will raise a new queen who will mate with a lot of drones, changing the colony's genetics. Feeding also helps. EFB is a notifiable disease in many jurisdictions; contact your local authorities if you find EFB in your hives. Antibiotics are used for treatment in some countries. In Australia, antibiotics are generally not used in beekeeping because of strict regulations aimed at preventing antibiotic resistance and ensuring honey safety. Instead, beekeepers focus on biosecurity measures, good hive management, and maintaining strong, healthy bee colonies to prevent and manage diseases.

However, in other jurisdictions, it's more common for bees to be treated with antibiotics. It's important that strict guidelines be followed as antibiotics can leave residues that may last in the hive for years and could contaminate any honey produced. In extreme cases, it may be necessary to destroy the infected colony.

Prevention: Ensure bees have plentiful access to nectar and pollen, and supplement with sugar syrup and/or pollen patties if needed. Order queens with hygienic traits if possible. Maintain hygienic beekeeping practices. Stress is a key risk factor for bees, so take precautions to minimise this if moving hives.

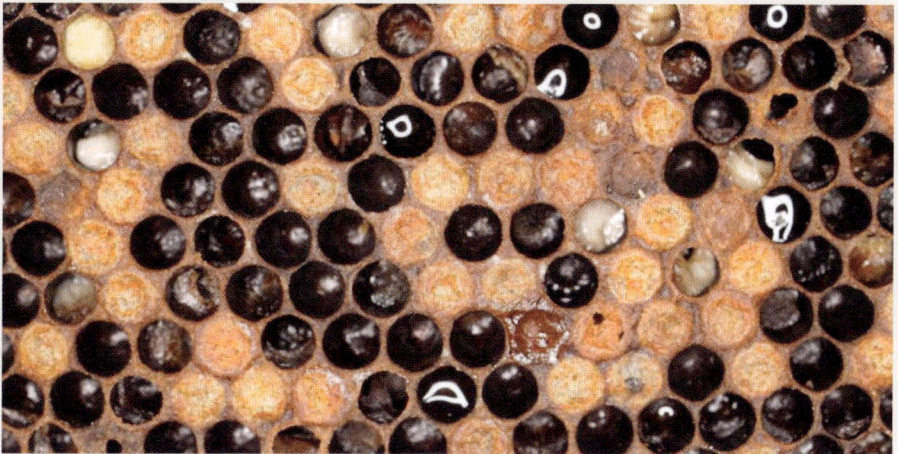

Dead watery larvae, an uneven brood pattern and a sour odour can indicate EFB.

Chalkbrood

Chalkbrood is a fungal disease that infects the gut of bee larvae. It is caused by a spore-forming fungus called *Ascosphaera apis*.

Symptoms: A chalky-white covering on larvae; grey or black fungus at the later stage of infection; mummified larvae in the tray or near the hive entrance.

Treatment: Moving the hive into full sun and increasing ventilation can help your bees to get on top of the fungus. A variety of pollen and a good nectar flow may enable your bees to beat chalkbrood. To assist the recovery, remove any mummies (chalky dead larvae) and any heavily affected comb. Try to avoid dropping mummies on the ground near your hives as it may further spread the fungus.

If the above methods fail, try requeening the colony. In some regions where this condition is prevalent, queen breeders select for chalkbrood-resistant genetics. Some beekeepers recommend burning any wooden frames that have been affected and replacing them with new ones.

Prevention: Position hives in full sun with adequate ventilation. Ensure adequate nutrition.

Larval mummification caused by chalkbrood

A bee with deformed wing virus.

Deformed wing virus (DWV)

One of the most common viruses affecting honey bees, deformed wing virus is primarily spread by varroa mites. (At the time of writing, DWV is not widely present in Australia.)

Symptoms: Deformed wings, shortened, rounded abdomens, discoloured legs and wings. Eviction of bees from the hive.

Treatment: Keeping varroa levels under control is important (page 244). If the colony appears to be failing, then requeening is advised, as both resistance and hygienic behaviour relating to DWV are genetically determined.

Nosema

Nosema is a common spore-forming fungal parasite that affects honey bees. Although usually benign, it can cause significant harm to colonies if a severe case is not addressed. There are two species of Nosema: *Nosema cerana* and *N. apis*.

Symptoms: Nosema infections often go unnoticed because there are no apparent signs of the disease. Look out for the following: Bees defecating inside the hive, and outside very close to the hive, leaving yellowish excrement stains on the top bars of frames, bottom board, combs and the inside and outside of the hive; signs of dysentery; poor attention to brood rearing, with bees turning to guard and foraging duties instead; inability to produce royal jelly; sick bees with distended abdomens crawling in front of the beehives with their wings spread out; failure to fly. If the queen is infected, egg production may drop and eggs may not produce larvae.

Treatment: Disinfecting contaminated combs is necessary to prevent further spread of the parasite within the hive. This can be achieved by fumigation with 60–80 per cent acetic acid vapour. The vapour kills the spores within one week.

Antibiotic treatments are available; however, in some jurisdictions, using antibiotics for Nosema treatment without a special permit is illegal.

Prevention: Maintain colony health and ensure the bees have adequate nutrition. Position the hive in a sunny winter location, preventing excess humidity within the hive. Avoid moving or opening the hive during winter.

Sacbrood

Sacbrood is a virus that affects honey bee larvae, and is most frequently found during phases of colony growth.

Symptoms: Larvae die off before fully transforming into pupae, leaving a distinct sac-like appearance in the affected brood; dying or dead larvae and pin-sized holes in cells; grey to yellow-brown discolouration. Larvae usually die upright, with the pointy end of the larva facing upwards.

Treatment: In most cases, the hive itself can overcome the presence of the disease through no additional steps. Locate the hive in full sun and increase ventilation. In severe cases, consider requeening the hive or relocating the entire colony to a non-infected site.

Prevention: Follow best-practice hive management techniques for a robust and well-nourished hive, and follow hygienic beekeeping practices as described on page 231.

Comb issues

Cross comb

Cross comb occurs when bees build brood comb across two or more brood frames, joining them together. This is common when foundationless frames are used but it can also happen with foundation sheets. When this happens, you can use a knife or hive tool to cut the comb that is wandering off the line and gently push it in line with the frame. The bees will quickly reshape the damaged or distorted comb.

If a piece of comb breaks off you can use rubber bands to temporarily hold it in place while the bees rejoin the comb. There is no need to remove the rubber bands as the bees will chew them away over time. In severe cases of cross comb, you may need to cut all the comb out of the frames. Cut the comb into sections that will fit within a frame, and then secure the sections with rubber bands.

Place your frames back in the brood box. Push the frames together, leaving any excess space on each side of the hive. Make sure there is at least a 4 millimetre (3/16 inch) gap between comb surfaces for the bees to service the comb. If there is not enough space between comb surfaces, small hive beetles could take advantage and lay in this area, leading to a slime out.

If necessary, rearrange the frames to allow bees to access all comb faces or cut away protruding comb to allow space.

If the cross comb is severe and most of the frames are joined together, it can be difficult to lift any frames without making a huge sticky mess. In this case, it is best to tip the whole eight or ten frames out of the brood box in one stuck-together mass as follows.

1   Set up a table near your hive and put the brood box onto it. With a long knife, cut any burr comb sticking to the wall and try to loosen the frames from their rest on the box wall.

2   Now, having made sure the queen is not on top of the frames, tip the whole box upside down onto the table. The idea is that the box itself can be lifted away, leaving the interconnected frames upside down on the table. If the box won't come off, you might have to slide one end of it over the edge of the table so that you can use a hive tool to loosen the propolis that's sticking the frame ends to the box.

3   Once you have lifted the box away from the combs it will be easier to tease each frame apart (watching out for the queen) and cut the sections of comb out so they can be reinserted into the frames using string or rubber bands.

Check the hive regularly. Once the bees get a few nice straight frames they will usually follow suit with the rest.

Combs have been gnawed

In colder climates mice might sometimes nest in your hive. If you notice nibbled comb, one or more mice may have moved into the hive. Remove mice if any are present and reduce the entrance size to prevent re-entry.

Presence of silky web and cocoons

Silky webs and cocoons might mean an infestation of wax moth or spider webs. If it's wax moths, turn to page 242. If it's spiders, you can simply remove them and their webs. If you are seeing webs, though, it means that your hive is probably weak for some reason and you will need to inspect your colony to find the issue.

Slime in the hive and/or on comb

This usually means small hive beetles. See page 240.

Larvae/maggots in comb (not bee larvae)

This is likely to be small hive beetle (page 240) or wax moth larvae (page 242).

Using elastic bands to repair cross comb.

Pests

It's highly likely that you'll have to help your bees deal with pests at some stage. Here are our tips on how to recognise and combat unwanted guests in the hive.

Ants (family Formicidae)

Ants are attracted to beehives as they provide warmth, shelter and a food source. Most ants won't cause a problem for your bees, as bees are good at keeping them out of the hive; however, they can be an annoyance to the beekeeper, creating a mess behind the window covers and under the roof. An extremely weak colony may struggle to keep ants out of the hive. In some places there are species of ants, such as Argentine ants, that can be very harmful to honey bee colonies.

Treatment and prevention: Brush away ants and their eggs from behind window covers and the key access cover. Remove them from under the roof too. Then create an ant barrier between the ground and the hive by using our ant guards filled with cooking oil or petroleum jelly or place the legs of your hive in a small container of vegetable oil or grease. Be sure to remove any foliage touching the hive that ants could use as a bridge.

Applying cinnamon powder behind the window covers and on top of the inner cover beneath the roof is another method to help to deter ants. Repeat the process a couple of times and the ants usually go away, until next time.

Asian hornet (*Vespa velutina*)

What does it look like? Asian hornets are 2.5 centimetres (1 inch) long and their head is a light orange shade with brown antennae and a yellow-orange base. They have large cheeks and pronounced lower parts of the face, with an orange lower jaw and a black tooth used for digging.

Where is it found? They are native to South-East Asia and are now present in most of Europe. *Vespa mandarinia*, the Asian giant hornet (also called the Japanese giant hornet), has been found in the Pacific Northwest of the USA, but at the time of writing it has not been sighted there for two years.

What does it do? These hornets prey on honey bees, which are very vulnerable to their attacks as they have not evolved natural defences.

Treatment and prevention: Reduce the size of the hive entrance using our entrance reducer or any makeshift solution. Install a hornet trap in or near your hive. Identify nest locations and check local regulations to see if they should be reported to environmental bodies; destroy their nests or hire a professional pest control company to do so.

A small hive beetle on drone brood.

A weak hive can be 'slimed out' by an abundance of small hive beetle larvae.

Small hive beetle (*Aethina tumida*)

What does it look like? Small hive beetles (SHB) are about 5 millimetres (³⁄₁₆ inch) long and dark brown or black.

Where is it found? Native to Africa, it has now spread widely in tropical, subtropical and warm temperate zones. In these places it's now normal to find SHB in your hive.

What does it do? A strong colony generally keeps small hive beetles in check with no intervention; however, a weak hive or one that has suffered a lot of damage to the comb structure will be vulnerable to their attack. The beetle larvae damage brood combs and contaminate honey. If their numbers get too high, they can cause a hive to collapse. Colonies that are weak in population are particularly at risk from infestation, and if the small hive beetle maggots are allowed to proliferate this can prompt the remaining bees to abandon the hive, leaving a slimy foul-smelling mess behind.

Treatment: If you notice small hive beetle maggots in the pest management tray (not to be confused with wax moth larvae), and/or sections of comb that are taking on a wet slimy look with maggots, you will need to act fast to save the colony. Firstly cut out or remove all brood frames containing only honey and pollen stores and replace with empty brood frames. The remaining brood frames containing brood should be placed together centrally in the brood box. Remove any honey supers, including any containing Flow Frames. Place the Flow Frames and any brood frames with honey stores you wish to keep in a freezer for at least three days to kill the beetle eggs. If the honey in the Flow Frames is spoiled by SHB maggots you will need to clean them (see page 248 for information on cleaning Flow Frames). If the honey and pollen stores removed are slimy and spoiled, the comb will need to be discarded and the frames cleaned prior to reuse.

In severe cases of slime out where the colony can't be saved, remove the honey super, discard the infected comb and clean the frames. Freeze your Flow Frames for three days to kill any remaining hive beetle eggs.

Prevention: Beetle traps, grease cakes, nematodes, salt boxes and screened bottom boards (such as the ones included with a Flow Hive) can all be used to help control SHB. One oil trap design hangs between the upper rails of brood or honeycomb frames. If you are using the Super Lifter to lift a box containing these traps, remove them before the box is swung up, or the oil will spill in the hive.

While the first defence is a strong colony, if it becomes weak for some reason then it's a good idea to trap the beetles to assist your colony in controlling them. There are also many DIY and commercial beetle traps available. If you have our pest management tray, you can cover the bottom of it with cooking oil to create a beetle trap – when the beetles fall through the screened bottom board they will drown in the oil. Be sure to keep any bees out of the tray by properly fitting the vented cover.

A strong colony can also fall victim to small hive beetles if there are areas of comb that the bees can't service. This can happen if your hive has fallen over and the gap between the comb faces is less than 4 millimetres (3⁄16 inch). When doing your brood inspections pay special attention when replacing the frames to make sure that there is at least a 4 millimetre gap between all comb surfaces.

Note that SHB don't need to parasitise honey bees; they can also reproduce in rotting fruit. So you might want to consider moving your compost pile if it's close to your hives and the beetles are causing problems.

Tracheal mite (*Acarapis woodi*)

What is it? A parasite that lives and reproduces in the trachea of European honey bees.

Where is it found? Tracheal mites have been reported in North America, Europe and Asia.

What does it do? The mite clogs the breathing tubes of adult bees, blocking oxygen flow and ultimately killing them. Symptoms may include disjointed wings, dysentery and a tendency to swarm.

Treatment: Requeening is effective. Menthol pellets and grease cakes can also be used.

Prevention: Genetic resistance is common. Cleanliness and regular hive inspections are recommended.

Wax moth / Waxworm (*Galleria mellonella, Aphomia sociella* and *Achroia grisella*)

What does it look like? Wax moths have brown wings. Their larvae are usually 2–2.5 centimetres (roughly 1 inch) long and white or brown in colour.

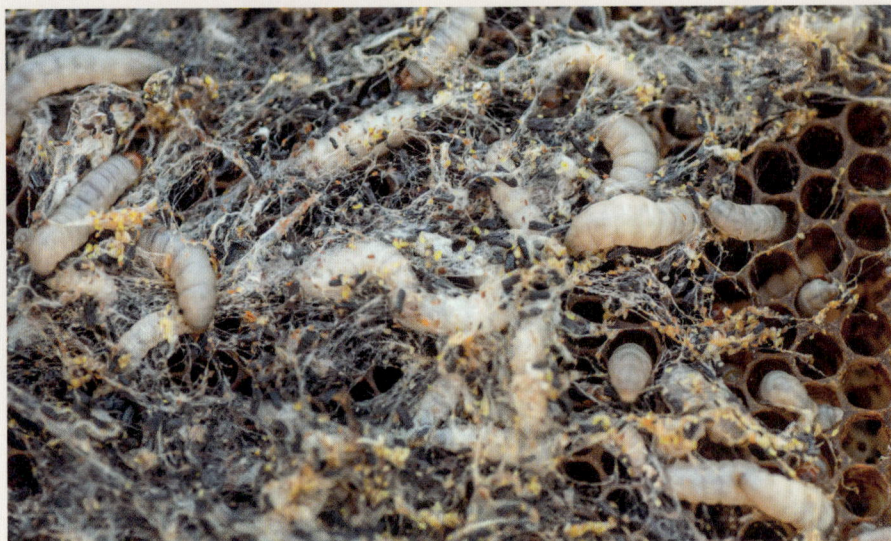

Wax moth larvae and their cocoons.

Where is it found? Virtually worldwide.

What does it do? Wax moth larvae feed on wax and produce a silky web that spreads across the hive into honeycombs and larval cells, making honey unusable. In the Flow Hive tray, you may see the large larvae of the wax moth. They eat the flecks of wax that fall down through the screened bottom board. They're most likely not a significant problem, because a strong hive will keep them out. To prevent them from increasing their numbers, it's beneficial to clean out your tray periodically; however, if you have a failing hive, the wax moth can take over. Left unchecked, wax moths can ruin the comb and chew away the woodwork. If their numbers build up you will notice the web-like tracks they make in the comb, and you may see their cocoons attached to wooden surfaces.

For conventional beekeepers, wax moths can be a real problem because they attack the stickies (honey supers that are stored after extraction before they are put back on a hive), whereas Flow Hive beekeepers don't have to keep a shed full of sticky honey frames. Even in a failing hive, if wax moths move in they can't damage the Flow Frames. So if you do find damage from wax moths in your Flow Hive, then it's likely that your colony is extremely weak, and you'll need to work out why.

Treatment: Remove moth larvae manually. DIY traps can be made using banana peels, sugar and vinegar. Freeze frames to kill any wax moths and their larvae and eggs. Remove and burn infested frames and containers as a last resort.

Prevention: Maintain a clean hive and a strong, healthy colony. If you are storing supers with wax comb in them over winter, take care not to allow moths to enter.

Tropilaelaps mite (*Tropilaelaps* spp.)

What is it? The original host of the Tropilaelaps mites is the giant honey bee species, *Apis dorsata*, which is native to Asia. Since successfully making the host shift from the giant honey bees to *A. mellifera*, there is a risk that Tropilaelaps will spread beyond Asia to other parts of the world.

What does it look like? The Tropilaelaps mite (called Tropi for short) is about a third the size of varroa: 1 millimetre (1⁄32 inch) long and 0.5–1 millimetre wide; however, they're still visible to the naked eye. When stationary they can be difficult to spot, but their elongated body makes them easier to see when they're moving.

Where is it found in a hive? Tropilaelaps' chelicerae (their fangs, loosely speaking) aren't strong enough to pierce through the outer layer of the adult honey bee, so they spend most of their time in the developing brood. Tropilaelaps can only survive a few days without brood. Once they have completed a reproductive cycle, they will usually enter another cell soon after. This limited period spent on the adult bees and ability to re-enter soon after emerging allows Tropilaelaps populations, and thus their effects, to grow exponentially by comparison with varroa mites populations.

What does it do? While there are currently four known Tropilaelaps species, only two (*T. clareae* and *T. mercedesae*) are currently known to successfully parasitise *Apis mellifera*. Unlike varroa, which creates a single feeding hole, Tropilaelaps creates numerous feeding holes all over the larva, causing the tissue to scar. As the bee develops, the places where the tissue has scarred can cause body abnormalities, such as bent antennae or legs. Tropilaelaps mite is also known to spread numerous viruses, including black queen cell virus and deformed wing virus (DWV).

Management: At this stage, there are few controls specifically designed for Tropilaelaps mites. The common detection methods include alcohol wash, sugar shake, the bump test and pressing sticky paper onto comb. Mechanical controls, including brood breaks by splitting, are common control practices across Asia.

Tropilaelaps' very short period outside of the capped brood makes them hard to treat. Treatments such as formic and oxalic acids and Amitraz have had varying levels of success. There are many things we don't yet understand about Tropilaelaps mite. As more research is completed, better identification and control methods will hopefully be developed.

Varroa mite (*Varroa destructor*)

Adult varroa mites are red–brown with a flat, oval-shaped body and eight legs.

Varroa levels should be monitored regularly. If mite loads exceed a certain threshold, your bees will be at risk if you don't act. Beekeepers around the world have been dealing with varroa mites for decades, and today there are a number of ways to effectively treat varroa. There's no single method that works best in all situations, so the ideal approach in your region will depend on a variety of factors. For more information and links to videos on managing varroa, see the Resources section on page 276.

What is it? *Varroa destructor* is an ectoparasitic mite that uses the adult honey bee and developing brood as its host. The varroa mite was originally a parasite of the Asian honey bee, *Apis cerana*. Varroa was first reported on *A. mellifera* colonies in the early 1950s. Honey bees haven't yet had much time to develop resistant genetic traits, and while there is great work being done to develop resistant breeds, in the meantime beekeepers need to give the bees a hand to deal with the mites.

What does it look like? Adult females are rusty red-brown in colour, with a flat, oval-shaped body and eight legs. Although tiny, they can be seen with the naked eye, especially against a light-coloured background.

Where is it found? Mites are now found on every continent where honey bees are kept, and are present in most hives in the world.

What does it do? The varroa mite is a highly destructive pest and a significant cause of hive losses. It feeds on brood and adult bees, weakening them and spreading debilitating viruses. If varroa levels are not managed there is a high probability that the colony will not survive.

Lifecycle: It's useful to understand a bit about varroa's lifecycle, as this helps to work out the timing of any interventions. The mite's lifecycle consists of two main phases: a reproductive phase and the time spent on adult bees (referred to as the phoretic phase). When a bee larva is ready to be capped, it emits an odour to notify the nurse bees. Varroa is able to identify this odour and enter the cell before the bees cap it. Sixty to 70 hours after capping, the foundress (mature female mite) will lay her first egg, which is always a male, before going on to lay females every 24–30 hours.

As the females reach sexual maturity, they mate with the male mite in the cell. At the end of a 21-day worker brood cycle, the bee will emerge with the foundress and, on average, 1.5 fresh female mites. (The male and any immature mites will die of dehydration once the cell is uncapped.) The lifecycle in drone cells is slightly different. Drones emerge from their cells after 24 days of development, with the foundress and an average of 2.5 additional mites. As the mites can breed at a higher rate in drone cells, and the drone larvae's bigger bodies provide more food, varroa has a strong preference for drone brood. As the infestation rate within a colony increases, it is common for numerous foundress mites to enter a single cell.

The second phase of varroa's lifecycle is the period spent on adult bees. This lasts for roughly three to fourteen days. The mites feed on adult bees (especially nurse bees) and use forager bees to disperse themselves to other locations.

Front view of an adult varroa mite.

Management: When dealing with varroa, it's not a matter of trying to eliminate the mites from your hive, but rather keeping their numbers below a certain threshold. This threshold varies depending on your climate and location, the time of year and what viruses are present in your region. We recommend consulting regional beekeeping organisations for guidance on treatment thresholds in your area (see the Resources section on page 276 for more information).

Monitoring: There are several ways to detect and count mites. One of the most common and effective methods is to 'wash' a sample of bees in alcohol to determine the bee:mite ratio in the hive. Sugar shakes and CO_2 anaesthesia are other methods that can be used to determine the mite load. Sticky boards are commonly used to count mites dropping through a screened bottom board, although this is far less accurate than an alcohol wash.

Treatment: Treatment methods are typically classified as cultural (affecting the mites' reproductive cycle), mechanical (removing mites) or chemical (poisoning the mites). Most beekeepers utilise a combination of these types of treatments at different times.

One common cultural control is to instigate an artificial brood break by temporarily caging the queen inside the hive so she's prevented from laying. This breaks the mites' reproductive cycle, as they reproduce on the developing larvae.

A popular mechanical control involves trapping mites in drone brood cells and destroying them before they emerge. Varroa prefer to reproduce on drone brood, so plastic foundation frames with cells sized for drone brood can be placed in the brood box towards the outer edges of the main brood area. The queen will tend to lay drone eggs in these larger cells, and the mites will preferentially lay in these cells and can then be culled by destroying the comb. This has to be done when the larvae are 14–23 days old (make sure not to wait longer than 23 days after adding the bait frame or all the mites could emerge into the hive).

A variety of chemical treatments are available, some organic (e.g. formic acid, oxalic acid, thymol) and some synthetic. Note that certain chemical treatments are prohibited or restricted in certain regions. Be sure to check your local regulations before applying any chemical treatments and follow all label directions and safety precautions. You may need to harvest or remove your honey super before applying a chemical treatment.

Breeding for resistance: Some breeders supply queens from genetic lines that are better able to deal with varroa than others. There are several traits that make bees more varroa resistant, including detecting and removing infested brood (also called 'varroa sensitive hygiene' or VSH), grooming themselves and each other to remove mites, and producing brood that develops in a shorter time.

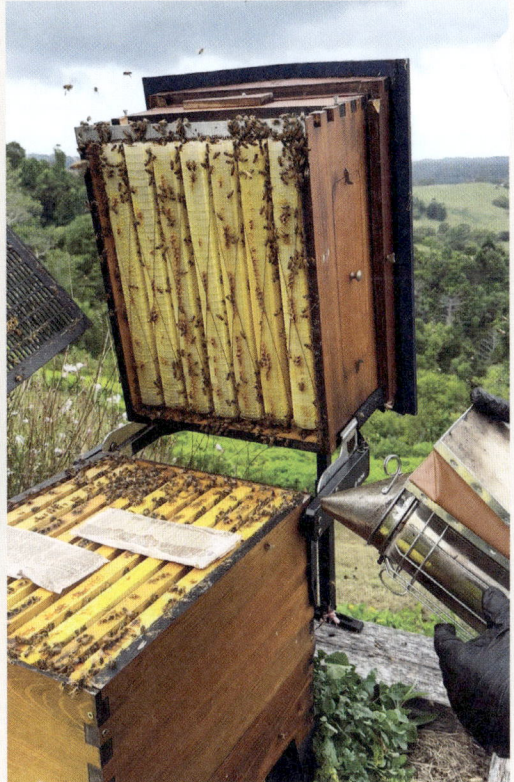

Clockwise from top A varroa mite on an adult honey bee; some treatment options for varroa involve adding strips to the brood box; an uncapped drone brood reveals the presence of a mite.

Flow Frame issues

Check this section if you've got any problems with your Flow Frames. We've also got lots of videos on our website: see the Resources section on page 276.

Maintenance and cleaning

Generally Flow Frames won't need regular cleaning, as the bees will keep the frames clean themselves (even if their idea of clean looks a little messy to humans). Bees are very hygienic insects and keep their hive in good order. However, over many years, the Flow Frames will become increasingly dark with wax, which can affect the rear observation window, so you may want to spruce it up from time to time. Sometimes mildew will grow on Flow Frames that have been stored outside the hive or if a colony has not been servicing them because it is weak.

We would like your Flow Frames to last as long as possible. If they are old and covered in dark wax, propolis and grime, please clean them and reuse rather than replace. We have many customers with Flow Frames that have been in use for over a decade now.

Some beekeepers soak Flow Frames in cleaning vinegar or bleach prior to washing. Washing should be done outdoors by leaning the frame against something so that it is supported. Although hot water sprayed from a hose can be effective, we have found using a pressure washer with hot water to be the most effective.

Set the Flow Frames to the open (harvest) position and remove the lower cap when cleaning so water can flow out effectively. Once you have removed most of the grime, allow them to dry thoroughly in the shade before storing them in a dark place or returning them to your hive.

While drying them outside for a day is fine, UV rays may damage Flow Frames, so they shouldn't be exposed to sunlight for extended periods.

Flow Frames spilling honey

It's not uncommon for some honey to spill inside the hive during harvesting with Flow Frames. It's also not uncommon to spill honey when harvesting in conventional ways, as prying off lids laden with honey and manipulating frames with burr comb often leads to a sticky mess for the bees to clean up. Bees get to work quickly mopping up honey and making repairs to the comb, so it rarely causes an issue.

Factors that can contribute to how the honey flows and whether there are any spills include differing viscosity throughout the frame, if bees are in cells, the slope of the hive, how far the bees have drawn out comb and capping beyond the Flow Frame surface, thixotropic or candied honey obstructing the flow, whether the cells were properly aligned prior to the bees filling them and temperature variations affecting viscosity.

Here are some tips to minimise any spills.

- Make sure your hive is on the correct honey harvesting angle so the honey can flow freely out and into your jar. There should be a 3 degree or more slope towards the harvesting end. Also make sure your hive is level side to side. We have found even experienced Flow Hive beekeepers can get this wrong, as the earth under your hive can sink. For this reason we've included spirit levels in the sides of our integrated hive stand – check them prior to harvesting.
- After harvesting it's important to properly reset the Flow Frames to the 'cell formed' position. We recommend leaving the key turned in the upper slot for a minute or two to make sure the cells are all returned to the correct position. If you do a quick close, the wax and propolis may create resistance and prevent the cells from fully aligning. This can lead to the bees using a lot of wax to create the cell shapes, which slows down the flow pathway next time you harvest, leading to spills. See below for more information on misaligned cells.
- Some customers find their honey is flowing so fast that the trough at the bottom overflows. This is a rare occurrence, but if it happens you can remedy it by harvesting the frame in segments. Simply insert the key part way and turn, then wait a few minutes before inserting it further and turning again.
- It's best to wait for the frame to be fully capped before harvesting so the honey is ripe and has a low moisture content for shelf life. Having the frame all or mostly capped can also minimise spills.

In most cases any honey spilt in the hive is easily cleaned up by the bees. However, if you do find spilt honey is causing issues, we suggest harvesting two or three frames and saving the rest for next time as this will lessen the issue.

Honey spilling outside the hive

If the corflute slider or pest tray is not in the right place when you harvest, you could end up with some honey dripping out of the hive.

Flow Frame cells are misaligned

If the moving portions of the cells in your Flow Frames have been misaligned for some time, the bees may use a lot of wax to create the cell shapes, which can then make it hard to realign the cells by turning the key. The extra wax can obstruct the flow of honey and cause spills when harvesting. If this has happened, remove the frame from the hive, then place a Flow Key (or ideally two keys) in the upper slot and turn 90 degrees to apply pressure. Put the frame in a black plastic bag or tub and leave it in the sun for a few hours. The wax and propolis will soften and the Flow Frame parts will properly align.

Alternatively, you can clean the wax and propolis off manually with a pressure washer before turning the key. The frame can then go back into the hive for the bees to tidy up.

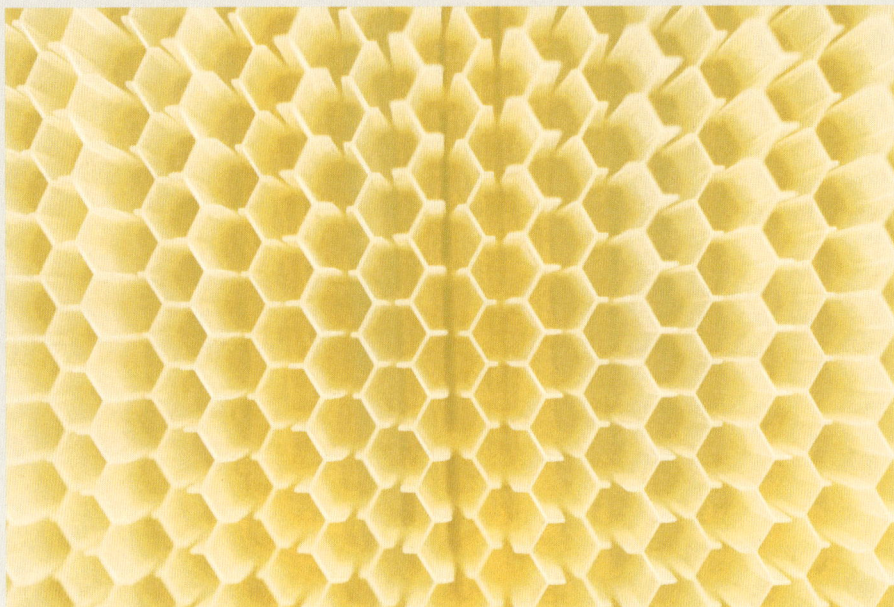

Make sure your Flow Frame cells are properly aligned so that they form hexagons.

I need to reassemble my Flow Frames, or adjust the tension

Our Flow Frames have been designed so that you can adjust the tension or reassemble them if they've come apart. See the videos linked in the Resources section on page 276.

The honey trough needs cleaning or the leak-back gap is blocked

When the tube is inserted for harvest, the 'tongue' on the honey tube should clear the leak-back gap. While this gap is clear, it will function to allow any honey that accumulates to go back to the bees. However, bees will be bees and block up everything, so sometimes it is necessary to manually clear this gap to stop honey building up in the area. This is easily done by using a matchstick, the end of your Flow Key or the harvesting tube. Before harvest, inspect the honey trough. If it is dirty, it can be cleaned from outside the hive using a clean wet cloth on the end of the key, or similar.

Honey has crystallised in the Flow Frames

If you're in an area where a lot of ivy, canola/rapeseed, sunflower or aster is growing, your honey might be prone to crystallising quickly. If this is the case, harvest as soon as the honey is ready, otherwise it may not flow readily.

However, if honey has crystallised in your Flow Frames, insert the Flow Key and move it to the harvest or open position. The key will be difficult to turn, but even a slight movement may crack the cells a little, prompting the bees to remove the crystallised honey and repair the comb. If that doesn't work you can heat the honey to decrystallise it by removing the Flow Frames from the super and putting them in an insulated box, such as a picnic cooler, with a heat source like a hot water bottle or an incubator heater. The frames must be warmed to 35–50 degrees Celsius (95–122 degrees Fahrenheit) for the honey to decrystallise. Return the Flow Frames to the super and harvest as usual.

Some beekeepers have also had success placing their super with crystallised honey underneath the brood box and then adding an empty super on top of the brood box. Bees prefer to store their honey above the brood, so they may move the honey to the empty super, decrystallising it in the process.

If you are nearing the end of the nectar season, leave the crystallised honey in the Flow Frames for the bees' winter supplies. (Remember that if you leave a box of honey on top of the brood in cold climates, you must remove the queen excluder to allow the queen to move into the super with the bees.)

The Flow Key won't turn

Over time, excess wax and propolis can build up in the Flow Frames, making the mechanism difficult to move, especially if the frames haven't been harvested for a long time. Insert the key a short distance into the Flow Frame (for example, just a quarter of the way) before turning it. This will reduce the force required as you will be operating only a portion of the Flow Frame. Leave the key in, with the frame under tension, for 30 seconds to a minute. This is an important step as it gives the wax a chance to come free. Then, insert the key further into the frame and turn it, again leaving it under tension for another 30 seconds to a minute. Repeat this process until the whole frame is in the open position and honey flows.

If this doesn't work, two opposing keys may be used together and turned at 90 degrees towards each other to produce more force. If the cells are still stuck, leave the two keys in position until the cells lift. In an extreme case the Flow Frames may need to be warmed to soften the wax and allow the parts to lift.

Honey issues

When you establish a new colony, it will take some time before the bees are ready to have a super added, depending on the strength of the hive, weather conditions and the availability of forage. It may take an entire season before the bees build up their numbers to the point that they're ready to move into a super, and after that it can take some time before they start storing surplus honey. You should only harvest when the bees won't be left short and when the honey in the super has been sufficiently capped. See page 151 for information on when to add a super, page 163 for details on when to harvest, and page 207 for feeding instructions if you think the colony's diet needs to be supplemented.

No honey in the hive

In order to get surplus honey you need a strong colony with lots of forager bees aligning with a good nectar flow that continues long enough for the colony to store honey beyond its own needs. If the hive is full of bees, there's been plenty of forage available and the weather hasn't been adverse for long periods, possible causes of a lack of honey being stored include the following.

- Health: pests or disease could be slowing the hive down.
- Genetics: bee colonies vary widely in how productive they are.
- Timing: perhaps your colony built up nicely with the nectar flow, but the flow stopped right when they were ready to start storing excess honey.
- The queen: your queen could be failing, or the hive could be queenless.

Honey has fermented in the jar

If the moisture content in your honey is too high, fermentation is likely to occur at some stage. For this reason, it's important to harvest after the bees have lowered the moisture content by fanning and then put their wax capping on top. If you have harvested too early, use the honey sooner rather than later, before fermentation occurs. Keeping it in the fridge will make it last longer.

Thixotropic honey

There are several flowering plants whose nectar produces thixotropic honey that is challenging to harvest, including manuka and jellybush (*Leptospermum scoparium* and *L. polygalifolium*) in Australia, ling or heather honey (*Calluna vulgaris*) in Europe and grapefruit honey (*Citrus paradisi*) in North America. Thixotropic honey stiffens into a paste or jelly, but if stirred vigorously, it will go runny again for a number of hours. We often see thixotropic honey coming out of the Flow Frames in jelly-like globules. However, if in high concentration, the honey may not flow out of Flow Frames and can also be difficult to extract with conventional methods. These include warming the frames in a heated room and inserting needles that stir the honey in the cells, then spinning the frames more quickly than in conventional extractors. Backyard beekeepers use rotating combs that rapidly insert their spikes into the comb cells to stir the honey into its more liquid form. While you could go to these lengths to extract a high concentration of thixotropic honey from your Flow Frames, most people find that it's by far easier to leave it for the bees and harvest thixotropic honey when it is in lower concentrations.

Flow Frames look full but there's not much honey when I harvest

Sometimes, when forage is scarce, the colony will eat the honey stores in the super, starting directly above the brood nest where it is easily accessed. This can create an empty half-moon shape, with honey left at the end of the Flow Frames so they continue to appear full. A quick heft of the hive will give you an idea of how much honey is stored. You can also pull out the Flow Frames and visually inspect them for the most accurate picture.

Looking in the observation windows over time will give you a great idea of whether the bees are storing more honey or consuming their stores. When the bees are hungry, it's usually best to leave the honey for them. Or you might be able to harvest one of the edge frames, as the honey is probably being consumed right above the brood nest.

Moving a hive

You shouldn't be afraid to move a hive if it's not where you want it to be, but it's best not to move a hive if you don't need to.

Moving a hive a very short distance

Bees have an extraordinary navigation system that is so precise that even if you move a hive as little as 3 metres (10 feet), you will probably find all the forager bees returning to the old location.

If you wish to move your hive across your yard, you can either move it incrementally 2 metres (6 feet) a day to the new location or move it in one go using what we call the reorientation technique. This involves putting a leafy branch in front of the entrance after moving the hive. When the exiting foragers meet this obstacle, they will adjust their navigation system and around 95 per cent of them will reorient to the new location. The other five per cent will go back to the old spot, so if you want to, you can put a box there for the bees to gather in and periodically ferry them to the new location. Or if there are other hives nearby, there's no need to leave a box, as the remaining foragers will probably join those colonies.

Preparing for a longer move

First, think through the process and make a step-by-step plan. Always wear a full bee suit and have a smoker at the ready throughout the process. Take care to keep every-thing secure as you go. It's usually best to have somebody helping you (hives are heavy!).

Secure the baseboard and inner cover to the hive with a strap. It's best to remove the gabled roof. If it's a long-distance move (requiring a transport vehicle), we recommend using ratchet or tie-down straps or a metal hive strap (known as an emlock) to secure the boxes. You'll need to seal the hive entrance so that the bees stay inside during the move. The best time to seal the hive is at night or early morning when the bees are all still inside. If you see any stray bees outside the hive, you can use some gentle puffs of smoke to encourage them back in. For longer moves, you need to make sure that there's sufficient airflow during the journey – the mesh bottom in a Flow Hive will provide the required ventilation if you take out the tray or corflute slider. Cover the entrance with one of our handy entrance reducers (a piece of mesh or steel wool can work well otherwise), and block the hole in the inner cover. It is okay to leave the hive shut overnight for a move in the morning, but please avoid sealing it for long on a hot day.

For longer-distance moves, use a pick-up truck, ute or trailer (moving a hive inside a car is not recommended). Ensure that the hives are level and stable and strap them to the vehicle securely so they won't move during the journey.

Medium distance moves – up to 6 kilometres (3.7 miles)

After closing the hive at night, you can move it to its new location, but before opening the entrance, create a 'landmark' for the bees so they will reorient to the new position. This could be as simple as leaning several leafy branches against the hive's entrance. The purpose is to make the hive look completely different, thus prompting the bees to reorient to the new location.

Some bees may still fly back to the original hive site. If this happens, you can leave a box at that site and, in a day or two, return them to the new site. Do this a few times until they have reoriented themselves.

Another option is to move them twice: first to a faraway location for two to three days or longer, and then to their final destination. This way, they won't remember where they were initially located and will reorient to their new home.

Long-distance moves – 6 kilometres (3.7 miles) and further

Simply move the bees and once you've set them up in their new location, open the entrance and let them get acquainted with their new environment. The bees should reset their internal navigation system since it's outside the primary foraging area of their old location.

We once moved a few hives from Cedar's house to his brother Chris's place in the same valley. With at least 20 kilometres of winding road between the sites, we assumed they were at least 10 kilometres apart. However, a day after we moved the hives, hundreds of bees were flying around back at Cedar's place. When he looked at the actual 'beeline' distance, it was only 4.9 kilometres. Those forager bees had gotten up in the morning, flown up the valley 2 or 3 kilometres looking for nectar and found familiar landmarks. This triggered them to use their 'old' internal map and they'd headed back to their original home.

MEET THE BEEKEEPER

Learning with bees

Melvin, Fisayo, Ofeijiro, Leo and Nathaniel,
New Jersey, USA

Having grown up on a farm, Melvin always wanted to care for animals and produce his own food. He and his wife Fisayo love that their son Leo is learning about nature through bees. To see more of their story, check out the Resources section on page 276.

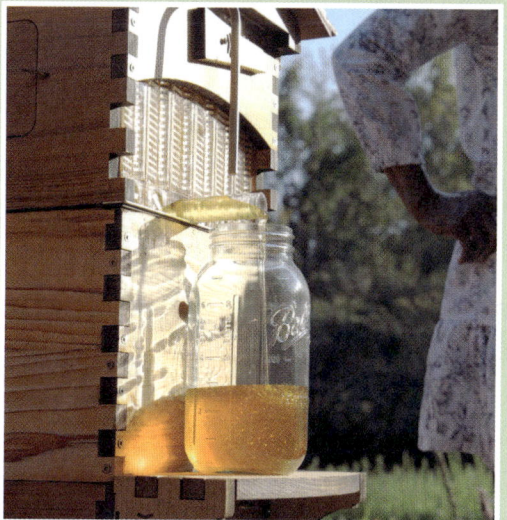

'When we first started, he had to literally drag me over. Now I'm just completely in awe of the bees. I think about them like the way I think about my kids.'

Scan here for videos relating to this chapter

9

Impact

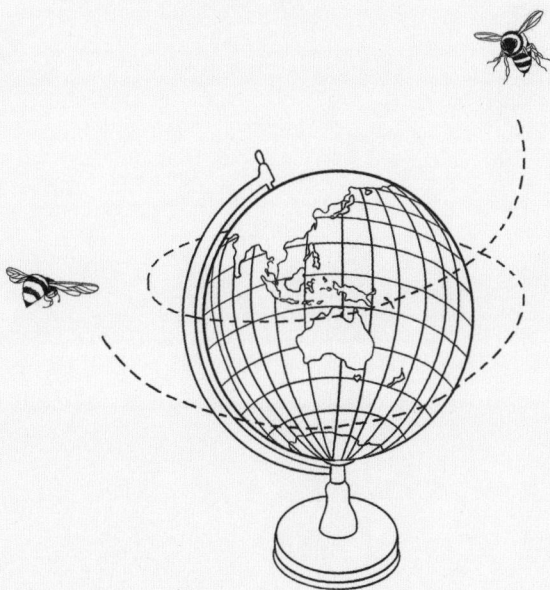

If we look after the bees,
they'll look after us.

For most of our lives, we'd been a bit cynical about the world of business. It seemed that an insatiable thirst for profit was the main thing destroying the natural world. While we still believe this to be true, as the journey with Flow unfolded, we increasingly began to understand the impact we could have through our company on issues we care deeply about. We were meeting many entrepreneurs who were giving back in a multitude of ways. We too wanted to join the ranks of the countless businesses that are making a positive impact.

For us, the cause that's always been closest to our hearts is protecting nature. Our existence is only possible because of the vast array of connections between living things. The air we breathe and the food we eat come to us only as a result of relationships between myriad plants and animals. And right now, on our watch, the sixth mass extinction event in Earth's history is happening. Some of the last remaining pockets of wilderness are being destroyed at a horrendous pace. These are biospheres that took millions of years to evolve: they won't come back if they're lost.

Being surrounded by so much phenomenally beautiful nature where we live, it's always felt obvious to us both that we should do whatever we can to help the web of life flourish. Once Flow got going, we could see that many of our customers cared a lot about the natural world, too. Bees tend to have that effect on people: if you spend time looking after them, there's a good chance you'll start to understand more deeply the way that these little beings are part of a bigger ecosystem, and that the overall health of that system affects *everything* within it.

You'll notice this interconnectedness when you watch bees fly out of a hive in the morning and later return with pollen. Or when new buds start to emerge in spring, and it makes you think about how busy your bees will be as they get ready for a big nectar flow.

And it's not just greenies like us who feel inspired by our winged little friends, but also everyone we talk to, not to mention our thousands of customers. It's so cool that no matter where we come from or how we vote or even what different life choices we make, pretty much everyone is on the same page when it comes to one thing: we must help our pollinators.

But this is just the start. The more humans recognise their connection to nature, the more likely it is that we'll make decisions – in areas ranging from households to global politics – that consider the vital importance of the interconnectedness of life. When we start to truly value where food comes from, the simple act of sharing a jar of honey or witnessing the magic of collecting nectar becomes more than just a positive experience: it is a transformative act with far-reaching consequences.

When we started out, we thought we were just creating a new and easier way to harvest honey. Hopefully we'd end up with enough money to live on and maybe have some left over to donate to a good cause. While we did achieve this, we gained an even deeper understanding of the significance of bees to our environment, and a greater appreciation for our community's willingness to support a cause they believed in. This led us to start thinking about how a business could have a bigger and more positive impact on the world. And this is what motivates us now.

Why the Flow Pollinator House?

We mentioned in a previous chapter (see page 27) that we raffled custom-built Flow Hives to raise funds for disaster relief in Vanuatu and Nepal. While these raffles were great, it felt like there was a limit to how much this approach could work over time. We wanted to develop a sustainable way of generating more funds so that we'd be able to have a bigger impact. And we wanted to focus our efforts primarily on projects that were geared towards helping bees and pollinators. This was the idea behind creating the Flow Pollinator House in 2017.

Although we didn't necessarily want to create another 'thing' or product to sell, we made a conscious choice to learn from everything we had already done and use it to our advantage. From our crash course in manufacturing of the Flow Hives, we figured out a way of reducing our production waste by recycling any timber offcuts into boxes that could be used to provide habitats for native bees and, at the same time, raise money to help the bees. This was a win-win situation. Not only would we be getting closer to our goal of zero waste in our production (something truly important to us, even though it is, at times, almost impossible), but we were providing more habitats for native bees and using the money generated for other projects designed to help them further. These projects included scientific research, sustainable development organisations with a beekeeping focus, citizen science initiatives and urban gardening for habitats.

The Flow Pollinator House provides a home for solitary nesting bees.

The Flow Pollinator House

These are made in our factory using timber offcuts from our hives and sustainably sourced bamboo. They're intended to provide a place for solitary native bees to take up residence. The best location for a pollinator house is an area protected from wind and away from tall weeds. It should be placed at least 1 metre off the ground but no higher than 2 metres. Native bees love gentle sunlight, so choose a bright area that's already attracting insects such as a flowering garden bed or an area with trees or a pond. Not only do pollinator houses offer a home to native bees, they can also become part of local pollinator corridors, providing a stepping stone for pollinators travelling between wild spaces.

Clockwise from top left *Tarsipes rostratus*, also known as the honey possum or noolbenger, is the world's only nectar-eating marsupial; the critically endangered swift parrot (*Lathamus discolor*) helps pollinate the world's tallest flowering plant, *Eucalyptus regnans*, in Tasmania; a monarch butterfly (*Danaus plexippus*); a pollinating checkered beetle from the family Cleridae.

Why do pollinators need help?

Everybody knows that pollinators are vital for food security and global environmental health. More than 75 per cent of the world's food crops partially rely on pollination by insects and other animals. And almost 90 per cent of wild flowering plants rely in part on animal pollination. Collectively, however, we don't seem to be very good at looking out for our pollinators. Massive-scale industrial agriculture soaks the land with insecticides and doesn't provide the variety of flower types that bees need. Deforestation and land clearing destroy their habitats. Add climate change, pests, diseases and air pollution to the mix, and you start to get an idea of what they're up against.

It's terrible to see a beehive that has died from exposure to pesticides. The ground around the hive is usually covered in dead bees, all with their little tongues sticking out. This is most likely caused by an unsuspecting neighbour spraying an insecticide on flowers in the daytime when the bees are foraging.

We think of the European honey bee as a 'gateway species'; they're cuter than most other insects, they give us honey and they pollinate our food. This makes people more inclined to love them, and that feeling often begins to expand to include the beauty and importance of all the other tiny living things that keep the natural world running.

Even though honey bees face many threats and a great number of colonies die each year, beekeepers will do everything possible to make sure these particular pollinators survive because the honey bee is so intertwined with the agricultural industry. This is despite the fact that a broad diversity of wild pollinators contributes to increased stability in pollination, even when managed bees, such as honey bees, are present in high numbers. Crop yields depend on both wild and managed species. So you can see why it's the native pollinators that need our help most of all. It's estimated that over 40 per cent of invertebrate pollinator species, especially bees and butterflies, are threatened in many regions. They are the ones who do the most pollination and provide vital ecosystem services such as pollinating crops, recycling nutrients into soils and decomposing waste. Researchers have declared: 'Without insects, our planet will not be able to survive.'

Where we live, we see an amazing assortment of native species. There are neon cuckoo bees, blue banded bees, masked bees, metallic green carpenter bees, fire-tailed resin bees and lots more. We've also got all sorts of butterflies, moths, flying foxes (a type of fruit bat), rainbow lorikeets (a parrot), antechinus (a tiny marsupial) and lizards such as skinks and geckoes. All help to pollinate certain plant species. In addition to food crops, pollinators also contribute to crops that provide canola and palm oils for biofuels, cotton, medicines, livestock feed and construction materials. Some species also provide materials such as beeswax for candles and musical instruments, and other arts and crafts.

When you think about this multiplicity of living things that all depend on each other to create such a glorious profusion of life, the link between pollinator health and a thriving biodiverse ecosystem becomes very clear. No wonder bees are considered a barometer of environmental world health.

Learning so much about bees and how essential they are to life on Earth caused us to decide to put aside our imagined limitations on what we could do as a company to help them. It was important to us and our team that we incorporated what we'd learnt into Flow's vision: the company exists to help bring about *a world in which pollinators are protected and celebrated for the key role they play in sustaining life on our planet'*.

Certification as a B Corp

We've always tried our best to tread lightly. Not consuming much and being reasonably self-sufficient has just been part of our way of life. We always look for clever ways to fix things or reuse them, hence installing the solar system and water turbine at the farm, and the trucks running on chip oil and things like that.

When our business grew, we were very mindful that its ecological footprint would get bigger, too. We always make sure that a close eye is kept on our operations and that negative impacts are minimised everywhere they can be. As well as the environment, there are many other things to think about, such as making sure that everybody involved in the supply chain is treated well.

Becoming a B Corp certified business in 2018 has helped in keeping us accountable on all of these levels. It takes quite a bit of effort to go through the certification process (thanks to our general manager, Summer!), but we're proud to be part of a community of like-minded companies. We want to help promote the idea that committing to integrity and transparency is a crucial part of the way forward for all businesses.

The B Corp movement

B Corps (Benefit Corporations) are part of a global initiative to reshape business values by advocating for companies to prioritise social and environmental impacts alongside profits. To become certified, businesses undergo a stringent assessment of their impact, governance, transparency and accountability. Many adjust the legal framework they operate under in a way that obliges them to consider how their actions affect people and the planet as well as profit.

Engaging with communities, employees, suppliers and customers is central to the B Corp ethos. Ultimately, the initiative aims to demonstrate that profitability can coexist with making a positive difference in the world, driving systemic change towards a more inclusive and responsible economy.

TheBeekeeper.org

Once our first attempt at generating funds for impact through sales of the Flow Pollinator House had been successful, we decided to do it again.

Becoming a beekeeper can be a daunting undertaking. Although it's not a difficult pastime, it's good to school up a bit so that you know what you're doing and can give your bees the best care. There's lots of content available online these days, but you'll often find contradictory information and it's not always easy to get answers to questions that arise. Because so many people in our community were new to beekeeping, we wanted to make sure they could easily access the knowledge they needed to confidently get started. We started building an online educational platform for beginner beekeepers and decided to use half of the profits that came from the membership fees to help pollinators.

Our aim was to produce easy-to-understand videos teaching people how to keep bees, and also evoke a sense of wonder at how special these creatures are (for a sample, check out the link in the Resources section on page 276). In addition to recording our own input, we asked well-known beekeepers and many of the world's top bee scientists to share their knowledge and expertise. After a ton of work, we officially launched TheBeekeeper.org at the start of 2020.

Billions of Blossoms program

TheBeekeeper.org was well-received, so we had to figure out how best to utilise the funds raised in service of helping pollinators, particularly bees – both honey bees and native species. We sat down with our strategy manager, Niall Fahy, to research and ponder.

To protect pollinators, you need to protect plants. And because everything in nature is connected, to effectively protect plant life you have to protect entire ecosystems. The broader tapestry of life in a region has to be kept healthy.

As we are so busy with the company, we knew we would need to find good collaborators. There are so many people out there doing fantastic work in this sphere, and we're honoured to now be working with a number of carefully chosen partners: organisations both big and small, locally and internationally. Some of them are fighting to protect biodiverse habitats to preserve the 'living libraries' that have taken millions of years to form. Others are working with Indigenous communities to reforest areas that were previously cleared, one is providing resources to farmers to make their practices more pollinator-friendly and another is creating pollinator highways in urban landscapes. We're keen to deepen this work and find ways to make a bigger positive impact, particularly for native bees.

We've called this program 'Billions of Blossoms'. At the time of writing more than AU$1 million (US$661,100) has been given away to projects in eleven different countries. And there's more to come soon!

Above, from left Planting the right species will provide forage for native bees and other pollinators; wildflower meadow conservation benefits bees, butterflies and other species. **Opposite page** Protecting and restoring biodiverse ecosystems is vital for pollinator health.

What can you do?

Whether you're a beekeeper or not, there are lots of ways that you can help pollinators out. If you've got a garden, one of the most helpful things you can do is to stop using highly toxic pesticides, or to spray only in the evening when flowers have closed. There are other ways to deal with pests in the garden, such as companion planting and encouraging predator species. You can also let patches grow a bit wild so that there are more flowers and living spaces available. Lots of native plants are always a good idea, as is a supply of water. It's also helpful to support organic farmers, and you can even become active as an advocate for pollinator-friendly policies in your area. Go to the Resources section on page 276 for a link to our guide on helping the bees.

'To reverse the crisis requires reciprocity, giving more than we take. When reciprocity prevails, everyone benefits. When it's absent, injustice prevails. Regeneration means stitching together the broken strands that separate us from each other and the natural world.'

PAUL HAWKEN, American environmentalist and author

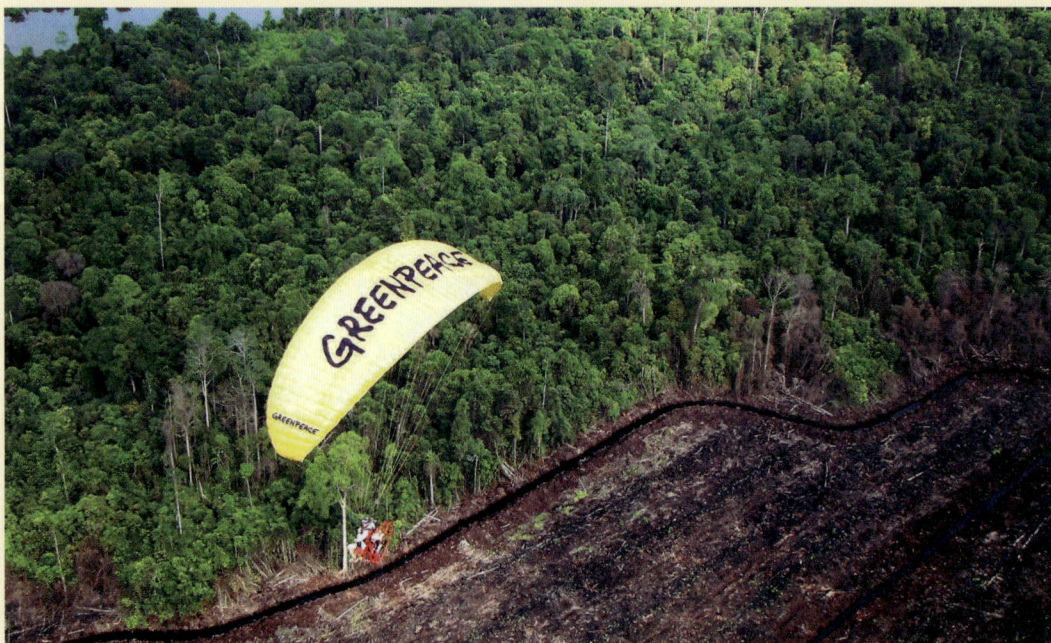

Cedar paragliding for Greenpeace in Sumatra. **Following pages** An Aussie favourite: the teddy bear bee (*Amegilla bombiformis*).

Looking forward

We think that the biggest positive impact we can make is through innovation. As well as trying to make beekeeping easier and gentler, we're often discussing inventions that could replace or adjust things that adversely affect the planet. Of course, 99 per cent of the ideas won't work out, but that 1 per cent just might have legs. Working together solving tricky problems is our happy place; we'll always be looking for a better way.

> 'You never change things by fighting the existing reality. To change something, build a new model that makes the existing model obsolete.'
>
> **RICHARD BUCKMINSTER FULLER, American designer, inventor, author and futurist**

This is a critical time in history. To us, the ecological crisis doesn't feel like something far away. Whole species are going extinct at an alarming rate. In the years since we launched the Flow Hive, we've personally experienced no fewer than three unprecedented climate-related natural disasters. The impetus to do whatever we can to keep the planet's life support systems running is as strong now as it's ever been.

Our journey with Flow has taught us both so much and it's helped to deepen our sense of purpose. We're still driven and optimistic, and we both feel incredibly fortunate that through Flow we can join forces with people who are doing fantastic things all around this beautiful world.

Despite the prevalence of bleak news everywhere these days, there have never been more people working together to come up with ways to make things better. For example, our understanding of soil carbon has increased dramatically in a couple of decades, opening practical possibilities of drawing down the excess carbon in the atmosphere. Just looking through the Billions of Blossoms projects is heartening, showcasing as they do a tiny fraction of all the wonderful people working for positive change.

While there are a lot of problems in the world to solve and it often seems as though solutions are out of reach, at the same time, we still think the first step towards creating something new is believing you can make it happen.

Our Beekeeping Family

It must be in our genes! Here are a few snaps of our clan and their bees.

Resources

Use this link to see videos with more detailed info on various topics discussed throughout the book:

honeyflow.com/book-resources

Or you can go straight to the online content for specific chapters by using the QR codes below:

Beginnings

1
Bees, the universe and everything

2
Beekeeping basics

3
Beehive basics

4
About the Flow Hive

5
Caring for your bees

6
The honey harvest

7
The beekeeping year

8
Hive health and troubleshooting

9
Impact

Meet the Beekeeper

References

Chapter 1

39 **There are roughly 20,000 species:** Orr MC, et al., 'Global patterns and drivers of bee distribution', *Current Biology*, 2020, Vol 31:3, 451–458. cell.com/current-biology/fulltext/S0960-9822(20)31596-7 accessed: 25/07/2025

41 **Each flower produces a unique:** Clarke D, Morley E, Robert D, 'The bee, the flower, and the electric field: electric ecology and aerial electroreception', *A Journal of Comparative Physiology: Neuroethology, sensory, neural, and behavioral physiology*, 2017, Vol 203:9, 737–748. doi: 10.1007/s00359-017-1176-6 accessed 25/07/2025

49 **Professor Lars Chittka, one of:** Solvi C, Baciadonna L, Chittka L, 'Unexpected rewards induce dopamine-dependent positive emotion-like state changes in bumblebees', *Science*, 2016, Vol 353:6307, 1529–1531. doi: 10.1126/science.aaf4454 accessed 25/07/2025

49 **'These unique minds, regardless of':** McGivney A, '"Bees are sentient": inside the stunning brains of nature's hardest workers', *The Guardian*, 2 April 2023. theguardian.com/environment/2023/apr/02/bees-intelligence-minds-pollination accessed 25/07/2025

50 **When we try to pick:** Muir J, *My First Summer in the Sierra*, (Houghton Mifflin, 1911) Sierra Club Books, 1988, p 110.

Chapter 2

57 **This rare insight into early:** Nayik G, et al., 'Honey: Its history and religious significance: a review', *Universal Journal of Pharmacy*, 2014, Vol 3, 5–8. researchgate.net/publication/260677542_Honey_Its_history_and_religious_significance_A_review accessed 25/07/2025

58 **In an intriguing example of:** Wood BM, et al., 'Mutualism and manipulation in Hadza–honeyguide interactions', *Evolution and Human Behavior*, 2014, Vol 35:6, 540–546. doi: 10.1016/j.evolhumbehav.2014.07.007 accessed 25/07/2025

63 **North America has around 4000:** Orr MC, et al., ibid.

63 **Often referred to as 'super:** Aizen MA, et al., 'How much does agriculture depend on pollinators? Lessons from long-term trends in crop production', *Annals of Botany*, 2009, Vol 103:9, 1579–1588, doi: 10.1093/aob/mcp076 accessed 26/07/2025

65 **a queen's development is 'reprogrammed':** Kamakura M, 'Royalactin induces queen differentiation in honeybees', *Nature*, 2011, Vol 473, 478–483. doi: 10.1038/nature10093 accessed 26/07/2025

65 **One of these pheromones is:** Knapp RA, et al., 'Environmentally responsive reproduction: neuroendocrine signalling and the evolution of eusociality', *Current Opinion in Insect Science*, 2022, Vol 53, 100951. doi: 10.1016/j.cois.2022.100951 accessed 26/07/2025

68 **a 2018 study by the:** Reina A, et al., 'Psychophysical laws and the superorganism', *Scientific Reports*, 2018, Vol 8:4387, 2018. doi: 10.1038/s41598-018-22616-y accessed 26/07/2025

Chapter 3

87 **In his 1853 volume *Langstroth's*:** Langstroth LL, *Langstroth's Hive and the Honey-Bee: The Classic Beekeeper's Manual*, Dover Publications, 2004, p 168.

98 **Native Americans figured out that:** Richardson, BW, 'On the anaesthetic properties of the *Lycoperdon proteus*, or common puff-ball', *Association Medical Journal*, 1853, s3-1, 479. doi: 10.1136/bmj.s3-1.22.479 accessed 26/07/2025

101 **Professor Thomas Seeley, who first:** Bakker, Karen, 'How to speak honeybee', noemamag.com/how-to-speak-honeybee/ accessed 12/08/2025

Chapter 4

111 **Beeswax is expensive for the:** Winston ML, *Biology of the Honey Bee*, Harvard University Press, 1987, p 36.

111 **They start by building a:** Karihaloo BJ, Zhang K, Wang J, 'Honeybee combs: how the circular cells transform into rounded hexagons', *Journal of the Royal Society Interface*, 2013, Vol 10:86, 20130299. doi: 10.1098/rsif.2013.0299 accessed 26/07/2025

Chapter 5

123 **These glands are spreading Nasanov:** Bortolotti L, Costa C, 'Chemical communication in the honey bee society' in Mucignat-Caretta C, ed, *Neurobiology of Chemical Communication*, CRC Press/Taylor & Francis, 2014, Chapter 5. ncbi.nlm.nih.gov/books/NBK200983/

145 **If two queens emerge at:** Butz VM, Dietz A, 'The mechanism of queen elimination in two-queen honey bee (*Apis mellifera* L.) colonies', *Journal of Apicultural Research*, 1994, Vol 33:2, 87–94. doi: 10.1080/00218839.1994.11100855 accessed 26/07/2025

Chapter 6

161 **It's been theorised that energy-:** Crittenden AN, 'The importance of honey consumption in human evolution', *Food and Foodways*, 2011, Vol 19:4, 257–273. doi: 10.1080/07409710.2011.630618 accessed 26/07/2025

161 **By around 9000 BC, 'honey:** Vidrih R, Hribar J, 'Mead: the oldest alcoholic beverage', in Kristbergsson K, Oliveira J, eds, *Traditional Food: Integrating Food Science and Engineering Knowledge Into the Food Chain*, Vol 10, Springer, 2016. doi: 10.1007/978-1-4899-7648-2_26 accessed 26/07/2025

162 **we asked the University of:** Grace E, et al., 'Sensory properties of yellow pea and macadamia honeys from conventional and flow hive extraction methods', *Journal of the Science of Food and Agriculture*, 2020, Vol 100, 2027–2034. doi: 10.1002/jsfa.10221 accessed 26/07/2025

176 **It was even said that:** Haarhoff TJ 'The Bees of Virgil', *Greece & Rome*, 1960, Vol 7:2, 155–170. jstor.org/stable/641548 accessed 26/07/2025

176 **Not only good for wounds:** Bobiş O, et al., 'Honey and diabetes: the importance of natural simple sugars in diet for preventing and treating different type of diabetes', *Oxidative Medicine and Cellular Longevity*, Vol 2018:4757893. doi: 10.1155/2018/4757893 accessed 26/07/2025

176 **Honey is a potent prebiotic:** Schell KR, et al., 'The potential of honey as a prebiotic food to re-engineer the gut microbiome toward a healthy state', *Frontiers in Nutrition*, 2022, Vol 9, 1–10. hdl.handle.net/10453/163622 accessed 26/07/2025

177 **Some researchers claim that the:** McInnes M, *Honey Sapiens: Human Cognition and Sugars – the Ugly, the Bad and the Good*, Hammersmith Books Limited, 2023, p 176.

177 **We certainly don't recommend that:** Jansen SA, et al., 'Grayanotoxin poisoning: "mad honey disease" and beyond', *Cardiovascular Toxicology*, 2012, Vol 12:3, 208–215. doi: 10.1007/s12012-012-9162-2 accessed 26/07/2025

180 **Studies have shown that propolis:** Shanahan M, et al., 'Thinking inside the box: restoring the propolis envelope facilitates honey bee social immunity', *PLoS ONE*, 2024, Vol 19:1, e0291744. doi: 10.1371/journal.pone.0291744 accessed 26/07/2025; Dalenberg H, et al., 'Propolis envelope promotes beneficial bacteria in the honey bee (*Apis mellifera*) mouthpart microbiome', *Insects*, 2020, Vol 11, 453. doi: 10.3390/insects11070453 accessed 26/07/2025

181   **It is also high in:** Zulhendri F, et al., 'Recent update on the anti-inflammatory activities of propolis', *Molecules*, 2022, Vol 27:23, 8473. doi: 10.3390/molecules27238473 accessed 26/07/2025; Zullkiflee N, Taha H, Usman A, 'Propolis: its role and efficacy in human health and diseases', *Molecules*, 2022, Vol 27:18, 6120. doi: 10.3390/molecules27186120 accessed 26/07/2025

181   **Early promising results of royal:** Sharif SN, Darsareh F, 'Effect of royal jelly on menopausal symptoms: a randomized placebo-controlled clinical trial', *Complementary Therapies in Clinical Practice*, 2019, Vol 37, 47–50. doi: 10.1016/j.ctcp.2019.08.006 accessed 26/07/2025; Oršolić N, Jazvinšćak Jembrek M, 'Royal jelly: biological action and health benefits', *International Journal of Molecular Sciences*, 2024, Vol 25:11, 6023. doi: 10.3390/ijms25116023 accessed 26/07/2025; Morita H, et al., 'Effect of royal jelly ingestion for six months on healthy volunteers', *Nutrition Journal*, 2012, Vol 11, 77. doi: 10.1186/1475-2891-11-77 accessed 26/07/2025

Chapter 7

207   **Bees prefer a CP% of:** Zheng B, Wu Z, Xu B, 'The effects of dietary protein levels on the population growth, performance, and physiology of honey bee workers during early spring', *Journal of Insect Science*, 2014, Vol 14:1, 191. doi:10.1093/jisesa/ieu053 accessed 26/07/2025

207   **As they process the nectar:** Riddle S, 'The chemistry of honey', *Bee Culture*, 25 July 2016. beeculture.com/the-chemistry-of-honey/ accessed 26/07/2025

211   **They'll even drink pure ethanol:** Black TE, et al., 'Effects of ethanol ingestion on aversive conditioning in honey bees (*Apis mellifera* L.)', *Journal of Comparative Psychology*, 2021, Vol 135:4, 559–567. doi: 10.1037/com0000296 accessed 26/07/2025

211   **... the worker bees detach:** Heinrich B, Esch H, 'Honeybee thermoregulation', *Science*, 1997, Vol 276, 1013. doi: 10.1126/science.276.5315.1013e accessed 26/07/2025

Chapter 9

265   **More than 75 per cent:** Aizen MA, et al., ibid.

265   **'It's estimated that over 40 :** IPBES. *The Assessment Report of the Intergovernmental Science-policy Platform on Biodiversity and Ecosystem Services on Pollinators, Pollination and Food Production.* Zenodo, 7 Dec 2016. doi: 10.5281/zenodo.3402857 accessed 26/07/2025

265   **Researchers have declared:** 'Without insects: Axel Hochkirch, quoted in Weston P, 'Number of species at risk of extinction doubles to 2 million, says study', *The Guardian*, 9 Nov 2023. theguardian.com/environment/2023/nov/08/species-at-risk-extinction-doubles-to-2-million-aoe accessed 26/07/2025

270   **'To reverse the crisis requires:** Hawken P, 'Regeneration can restore a broken world', TED talk, 2023. youtube/q-tutSkZ8k8 accessed 26/07/2025

271   **'You never change things by:** Richard Buckminster Fuller, quoted in Vance M, Deacon D, *Think Out of the Box*, Career Press, 2006, p 78.

Bibliography

Beekeeping instruction

The Practical Beekeeper, Michael Bush, X-Star Publishing Company, 2011

The ABC & XYZ of Bee Culture, Al Root & ER Root, Literary Licensing, LLC, 2013

The Hive and the Honey Bee, LL Langstroth; Joe M Graham, ed., Dadant & Sons, 2015

The World History of Beekeeping and Honey Hunting, Eva Crane, Routledge, 1999

The Australian Beekeeping Manual (3rd edition), Robert Owen, Exisle Publishing, 2023

The Beekeeper's Handbook (4th edition), Diana Sammataro & Alphonse Avitabile, Comstock Publishing, 2011

The wonder of bees

The Mind of a Bee, Lars Chittka, Princeton University Press, 2022

Honeybee Democracy, Thomas D Seeley, Princeton University Press, 2011

What a Bee Knows: Exploring the Thoughts, Memories, and Personalities of Bees, Stephen L Buchmann, Island Press, 2023

The Buzz about Bees: Biology of a Superorganism, Jürgen Tautz, Springer, 2008

Bees of Australia: A Photographic Exploration, James Dorey, CSIRO Publishing, 2018

Pollinators & conservation

The Diversity of Life, Edward O Wilson, Penguin Press, 2001

The Garden Jungle: or Gardening to Save the Planet, Dave Goulson, Vintage Arrow, 2020

The Forgotten Pollinators, Stephen L Buchmann, Island Press, 1997

Attracting Native Pollinators: Protecting North America's Bees and Butterflies, Xerces Society, Storey Publishing, 2011

Melbourne Pollinator Corridor Handbook, Emma Cutting, The Heart Gardening Project, 2021

The Ecology of Commerce (revised edition), Paul Hawken, Collins Business Essentials, 2010

Acknowledgements

We acknowledge all First Nations peoples, particularly the Bundjalung Nation, who are the traditional custodians of the lands on which we live, work and play.

It's been a big effort to create this book, and we want to express our heartfelt thanks to all of the following people.

- Niall Fahy for directing and carrying the project from start to end and providing tons of writing and creative input. There's no way this book would exist without him!
- Jane Morrow, Alex Payne, Virginia Birch, Sarah Odgers, Emily O'Neill, Samantha Miles, Melody Lord and everyone else at Murdoch Books.
- All of the amazing people who have been part of the Flow team, past and present, or who've played a part in the story. Flow wouldn't be what it is without you!
- Kylie Ezart and Michele Wainwright for their love, support and patience over the years.
- Michael Bush for his expert region-specific advice on numerous aspects of beekeeping.
- Frewoini Baume, Martin Fahy, Nell Cook and Pete Wilkins for their important contributions to the text.
- Jai Anderson, Mirabai Nicholson-McKellar, Callum Griffith, Ben Alexander and Frederick Dunn for their photography throughout these pages, and everyone who has kindly contributed additional photos.
- Our wonderful customers and supporters around the world!

Index

Published in 2026 by Murdoch Books, an imprint of Allen & Unwin

Murdoch Books Australia
Cammeraygal Country
83 Alexander Street
Crows Nest NSW 2065
Phone: +61 (0)2 8425 0100
murdochbooks.com.au
info@murdochbooks.com.au

For corporate orders and custom publishing,
contact our business development team at
salesenquiries@murdochbooks.com.au

Publishers: Alexandra Payne and Jane Morrow
Editorial manager: Virginia Birch
Design manager: Sarah Odgers
Design and illustrations: Emily O'Neill
Developmental editor: Justine Costigan
Additional writing and text development: Niall Fahy
Editors: Samantha Miles and Melody Lord
Food and lifestyle photographer: Rob Palmer
Production manager: Natalie Crouch

Photography credits: 40 (top left), 59, 242, 264 (top left)
© Alamy; 264 (top right) © Rob Blakers/Bob Brown
Foundation; 49 © Professor Lars Chittka; 269 (left) ©
Emma Cutting, Heartscapes; 244, 245 © Frederick Dunn;
Front cover image and 2x back cover images, 2, 6-7, 13,
14, 17, 18, 21, 25, 30, 32 (bottom left and right), 34, 35,
38, 40 (bottom left and right), 42-43, 44, 51 (bottom left
and right), 52, 53, 60 (bottom left), 64, 66, 69 (bottom left
and bottom right), 70-71, 75, 76, 77, 79, 80-81, 82-83,
86, 88 (top left and bottom left), 91, 92, 94, 99 (top right
and bottom right), 102, 104-105, 108, 112, 114, 115,
117, 122, 123, 124, 125, 126, 128-129, 132, 134 (bottom
left), 136-137, 138, 140-141, 143 (top right and middle
right), 144, 145, 147, 148-149, 153, 154, 156-157, 165,
167, 168, 170-171, 172 (top right, middle left), 174, 175,
181, 196, 199, 200-201, 203, 205, 206, 209, 210, 212,
215, 216-217, 218, 219, 225, 228, 229, 232, 233, 234,
237, 240, 247, 250, 255, 256, 257, 260, 263 (top left and
right), 264 (bottom left, bottom right), 267, 270, 272-273,
274-275 © Flow Hive; 269 (right) © Kósa István; 51 (top)
© Joe Neely; 4, 8-9, 28, 32 (top left), 33, 56, 60 (top left,
top right and bottom right), 62, 88 (right), 96, 99 (left),
116, 120, 127, 134 (top left and right), 143 (left and
bottom), 150, 160, 166, 169, 172 (top left, middle right
and bottom x2), 178, 182, 185, 186, 189, 190, 193, 238,
263 (bottom left), 281, 286-287 © Rob Palmer; 268 ©
ReForest Now.

Every reasonable effort has been made to trace the owners
of copyright materials in this book, but in some instances
this has proven impossible. The author(s) and publisher will
be glad to receive information leading to more complete
acknowledgements in subsequent printings of the book and
in the meantime extend their apologies for any omissions.

Murdoch Books UK
Ormond House
26-27 Boswell Street
London WC1N 3JZ
Phone: +44 (0) 20 8785 5995
murdochbooks.co.uk
info@murdochbooks.co.uk

*Murdoch Books acknowledges the Traditional Owners
of the Country on which we live and work. We pay our
respects to all Aboriginal and Torres Strait Islander Elders,
past and present.*

EU Authorised Representative: Easy Access System
Europe, Mustamäe tee 50, 10621 Tallinn, Estonia, gpsr.
requests@easproject.com

ISBN 978 1 76150 004 6

A catalogue record for this
book is available from the
National Library of Australia

A catalogue record for this book is available
from the British Library

Colour reproduction by Splitting Image Colour Studio Pty
Ltd, Wantirna, Victoria

Printed in China by 1010 Printing International Limited,
China

OVEN GUIDE: You may find cooking times vary
depending on the oven and oven setting you are
using. For fan-forced (convection) ovens, as a general
rule, set the oven temperature to 20°C (25–50°F)
lower than indicated in the recipe.

TABLESPOON MEASURES: We have used 20 ml
(4 teaspoon) tablespoon measures. If you are using
a 15 ml (3 teaspoon) tablespoon add an extra
teaspoon of the ingredient for each tablespoon
specified.

DISCLAIMER: The content presented in this book is meant
for inspiration and informational purposes only. The author
and publisher claim no responsibility to any person or
entity for any liability, loss, or damage caused or alleged to
be caused directly or indirectly as a result of the use,
application, or interpretation of the material in this book.

10 9 8 7 6 5 4 3 2 1

MIX
Paper | Supporting
responsible forestry
FSC® C016973